NOV 2 2 2017

P9-CFO-088

GAME CHANGE

GAME

THE LIFE AND DEATH OF STEVE MONTADOR

CHANGE

AND THE FUTURE OF HOCKEY

KEN DRYDEN

SIGNAL

McCLELLAND
& STEWART

Hardcover edition published 2017

Library and Archives Canada Cataloguing in Publication
is available upon request.

ISBN: 978-0-7710-2747-5
ebook ISBN: 978-0-7710-2748-2

Library of Congress Control Number is available upon request

Book design by Five Seventeen
Jacket image: Steve Montador #4 of the Buffalo Sabres listens to the National
Anthem before playing the San Jose Sharks on February 13, 2010 at HSBC Arena
in Buffalo, New York. (Photo by Bill Wippert/NHLI via Getty Images)
Typeset in Janson by M&S, Toronto
Printed and bound in the USA

Published by Signal,
an imprint of McClelland & Stewart,
a division of Penguin Random House Canada Limited,
a Penguin Random House Company
www.penguinrandomhouse.ca

1 2 3 4 5   21 20 19 18 17

Penguin
Random
House

SIGNAL
McCLELLAND
& STEWART

To Khaya, Hunter, Mara, and Blake,
and tomorrow's players everywhere

I never met Steve Montador. In his early seasons with the Calgary Flames, I was president of the Toronto Maple Leafs, so I must have seen him play—but I have no recollection of him.

I love sports. I played sports all my life. My son and daughter played hockey. My two grandsons play today. When I heard the news of Steve's death, I wanted to know how, and why, a thirty-five-year-old, recently retired hockey player had died.

This is Steve's story, and it's the story of a game—of where it began, how it got to be where it is, where it can go from here, and how it can get there.

GAME CHANGE

It was the morning of February 15, 2015, and Dr. Lili-Naz Hazrati, a neuropathologist at the University Health Network in Toronto, was listening to her radio when she heard the news that Steve Montador, a former NHL defenceman, had died. Hazrati studies the brains of people who have exhibited symptoms of brain disease or dysfunction in the months or years before their death—as a result of Alzheimer's perhaps, or Parkinson's, or amyotrophic lateral sclerosis (ALS), or chronic traumatic encephalopathy (CTE). Hazrati is not a hockey fan and did not know Montador's name. But as part of her work, she did know that in recent years CTE has been found in the post-mortem examinations of several former football and hockey players.

Born in Iran, Hazrati left with her parents and sister in the midst of the country's 1979 revolution, settling in Barcelona first, before coming to Canada when she was eighteen. After her first year in university during a summer job in a neurobiology research lab in Quebec City, she learned how to do experiments and surgical procedures, an experience that would set her on her career path.

Hazrati's interest is the brain, and she is now also a member of a sports concussion research group led by the renowned Toronto neurosurgeon Dr. Charles Tator.

To gather the necessary data for her research, Hazrati needs brains to study. So when she heard about Steve, she knew she had to act quickly—as his was a sudden and unexpected death, there would be an autopsy performed within twenty-four hours. Hazrati called the chief coroner and asked him if he would retain the brain, pending her request to the Montador family for their consent to the donation of their son's brain.

Two weeks later, Steve's brain arrived at her office at the Toronto General Hospital (TGH). It was in a cardboard box inside a plastic bag containing a formaldehyde solution called formalin, which helps to preserve the brain's tissue. Hazrati cut open the bag and put Steve's brain in a blue hairnet—the kind hospital workers wear—and placed it in a large white plastic bucket filled with fresh formalin. She then attached the ends of the hairnet to the sides of the bucket, to allow the brain to float in the solution. There was no outward sign of injury or disease, she noted—but then, there rarely is—and any difference in weight or configuration since the time of death would be slight. She did not examine the brain further. She labelled the bucket "A1545"; this was the hospital's forty-fifth autopsy case in 2015.

Steve's brain remained inside the bucket, in a storage area off the autopsy room, for two more weeks, "to fix properly," as Hazrati puts it—for it to become firm enough to hold its shape for examination and to be cut easily. After those two weeks, Hazrati opened the bucket containing Steve's brain and began her examination.

When Hazrati had first seen the brain, it was almost white, with spidery, pinkish-red blood vessels running along its surface. Now the brain was a grey-white colour, with those pinkish-red lines turned red-brown because of the formalin, and its contours more obvious. Hazrati weighed the brain and checked her reading against the one in the coroner's report: 1,600 grams (about 3.5 pounds), an increase of ten per cent. This was to be expected, because of the formalin that the brain had absorbed. Again, nothing was unusual.

On a normal day, Hazrati might have three or four brains to cut—from deaths following cardiac surgeries, from general autopsies, from neurodegenerative cases. But on that day in 2015, she had one: Steve's. She sees so few brains of "young ones," as she puts it—those who are forty years old or younger, but who during their lifetimes exhibited symptoms of those much older: memory loss, loss of emotional control, loss of cognitive function. Hazrati is rarely shocked by what she sees, because she never anticipates what she will find. She is a scientist, and she must see only what is there. "The organ is what gives you the most information," she says. "You must let it tell the story." Still, a young brain like Steve's that looked so normal and healthy—but wasn't—was a surprise to her. "This was a young person who shouldn't be dead now," she says. His brain should not have been on the autopsy table in front of her, in that small, spare, surgically clean room.

As Hazrati puts it, brains like Steve's are "very precious," and extra care must be taken. Not because they are from well-known athletes, but because these athletes have died so young—and because the disease these brains contain is so little understood that every glimpse into them matters. They are precious because, for researchers to discover what they haven't yet been able to, they need lots of brains in order to gauge differences, uncover similarities, and have new things to think about and new findings to share with other scientists. Researchers have examined brains affected by

Alzheimer's, Parkinson's, ALS, and other neurodegenerative diseases for decades. They still don't know the causes of these diseases, or how to cure them, but they have studied them enough to know there is a certain predictability to them. This is less so the case with CTE. It seems that the disease is related to blows to the head, but not limited to them. After all, even the most physically inactive among us have hit our heads hard many times in our lives. What is it about the brains of these athletes that are different?

Hazrati laid Steve's brain on the stainless steel table and began to cut. Her knife was long and razor-blade thin, sharp enough to slice through the full width and depth of the brain in one motion without tearing the tissue. Each cutting was about a half centimetre thick, a complete and smooth cross-section that captured all of the brain's structures and allowed Hazrati to compare one side to the other. She then used a scalpel to cut the tissue into even smaller segments. The hospital's technicians will embed these segments into wax to protect them and make them easier to handle, then put them into cassettes, and place them into a processor for a few days, in a bath of xylene, chlorine, alcohol, and formalin. Other organs require less time to hydrate and "fix," but a brain takes longer because of its high lipid or fat content.

For Hazrati, this day was about seeing that the cuts were done properly, and preparing them and ensuring that each was labelled accurately. There was nothing more that she could do at that moment; there was nothing more she could discover.

Several days later, the next stage of cuttings began. The cuts were much finer this time—using a machine not much different than a sausage slicer, only smaller—six microns in depth (about one-quarter the thickness of letter paper), and so thin that colour disappears from the cutting and everything becomes transparent to

4

the naked eye, as if nothing is there. The technicians put the new cuttings into another bath, of water this time, to take the wrinkles out of them, then removed them one at a time with a small painter's brush, placing each on a glass slide. A stain was applied to give definition to the almost-invisible specimen within, which allowed Hazrati "to see inside all those different cells, individually, one by one" with her microscope.

Depending on what she wants to test for, Hazrati can ask the technicians to apply different stains. The regular stain is pink, and with it she can pick up evidence of Parkinson's but not of Alzheimer's; the Alzheimer's proteins are too similar in colour to the stain, and so blend into the pink. Nor can she see CTE with the regular stain.

With Steve's brain, Hazrati was looking for abnormal proteins, not knowing what she might find: alpha-synuclein, which is evidence of Parkinson's; TDP-43, which indicates ALS; beta-amyloid, found in those with Alzheimer's; and ubiquitin, which is related to multiple diseases. Most particularly she was looking for tau, which is found in sufferers of Alzheimer's, but also in those with CTE. So, on the slides containing tissues from each section of Steve's brain, a technician also applied a brown stain. This stain contains an antibody that can "recognize" specific proteins such as tau—and, if tau is present, will "go and sit exactly where the tau is," as Hazrati puts it. The stain binds to the protein and "tags" it, showing the tau as distinctly brown.

Tau occurs naturally in the brain. Healthy tau allows different parts of the brain to communicate with each other, enabling the brain to function. This communication happens neuron to neuron, through long fibres called axons that connect together like the ends of the buckle on a child's car-seat harness. Tau is in the axons, and gives them the structural integrity they need to forge this connection.

Unhealthy tau causes the axons to lose their structural integrity, breaking the link between neurons, interrupting the communication between the brain's parts, and diminishing the brain's ability to function. Unhealthy tau, through a microscope, appears as clumps, like a pile of bricks after a house collapses. This is the signature evidence for CTE.

A technician placed a glass cover slip on top of each of the slides containing the samples from Steve's brain; the cutting was pressed smooth in between. Each slide was labelled by name and surgical number, and assigned a barcode, then put into trays that were delivered to Hazrati's office. She was working on a separate case when they arrived, and had other surgicals to get to. She decided to put the trays to one side and to examine them on the weekend.

On Saturday, March 21, Hazrati arrived in her office at 7:00 a.m. "What you hope for the day you examine your trays is that you're completely there," Hazrati says. That you are focused, in other words. "Because you can easily make mistakes. There's nothing that will jump out at you and tell you that it's there. It's up to you to see it. To recognize patterns. It's what our eyes have been trained to do." And if she does make mistakes—if she doesn't see what is there, or if she sees what isn't—it can misdirect her thinking and set her back weeks. It can affect the next researcher, too, in another case who is also working hard to solve the brain's puzzle, and the ones after that. Most of all, it can affect some future patient whose hope is in her hands.

Hazrati had other trays to examine that day, but she began with Steve's. And while most times she starts with her pink-stained slides and goes through all of them before moving on to the brown, this time she started with the brown slides. "I knew this was a young, healthy person and I'm not expecting to see unexpected things," she says. "So I went right to what nails the question: is tau there or not?" She began with a slide from the frontal lobe. If tau was present, it would show up here.

Divided into two hemispheres, the frontal lobe extends over the top of the head from the crown to the forehead and almost to each ear. In terms of the brain's function, this is where everything comes together. All of the sensory inputs gathered elsewhere in the brain come to the frontal lobe to be processed, and made sense of, and given meaning. This is where information gets stored in short-term memory, where it is examined and manipulated, where its problems are solved, where its complications are subjected to judgment, where emotions are added, where decisions get made. It is in the frontal lobe where our humanity resides—what makes humans *humans*. It is where our personality lies—what makes us *us*.

The temporal lobe begins where the frontal lobe ends, and extends on each side of the brain from the core of the eye to the back of the ear. It is the site of long-term memory. The parietal lobe covers the upper rear quadrant of the brain. It is a place of recognition and understanding—of a face, of language, of space and our orientation to it.

Hazrati had made cuts from each of these parts of the brain, and she examined every slide for anything that the stains might reveal. In Steve's case, the presence or absence of tau was *the* answer. But for Hazrati as a researcher, its presence or absence meant a different set of questions. If Steve didn't have tau, why not? Others with a similar history of concussions had it, so what was different about him? Genetics? The seriousness or frequency of the blows to his head? Or something else? Or maybe it would mean that she had missed something. Maybe something other than tau had caused his symptoms. And if tau were present, the questions would be the same, only reversed.

Hazrati knew that if Steve's brain showed signs of CTE, that was "amazing." If it didn't, it would be amazing for other reasons. In either event, she would be able to add these slides and the remaining block of Steve's brain tissue to those from every other brain she had worked on, to examine again later if some new brain revealed something that got her thinking in a new way. The answer, as she says, is in the organs. It is where all the information she needs to know lies. It is why Hazrati keeps looking through the eyepiece of her microscope, no matter what she finds—or does not find.

On that Saturday morning, Hazrati sat down at her microscope and pulled out her first slide, which contained matter taken from the frontal lobe. She put it on the glass, and through her eyepiece she was able to see inside the cell, into the neurons, the astrocytes. Later, she looked at slides from the temporal and parietal lobes. What she found gave her a "fuzzy feeling," as she puts it. She saw the brown that revealed tau.

Steve had CTE.

CHAPTER TWO

Training camp, September 2003. The Calgary Flames had missed the playoffs for seven straight years, and had finished the previous season seventeen points behind Edmonton, no less, for the last playoff spot in the Western Conference. With the low value of the Canadian dollar, no NHL salary cap, and a few thousand empty seats at most home games, Calgary, a city known for its optimism, was very pessimistic about the present and the future of the Flames. However, there was some promise. The year before, Darryl Sutter—one of the gritty, no-nonsense Sutters of Viking, Alberta—had been made coach midseason, and with him behind the bench the team had managed a slightly better than .500 record. There were also some young defencemen who were just emerging into their ready-to-contribute years—Robyn Regehr, Jordan Leopold, Toni Lydman, Denis Gauthier, Andrew Ference—and two others, less experienced, who were ready to push them—Mike Commodore and Steve Montador. Steve was twenty-three.

That summer, the Flames had also acquired Rhett Warrener, a defenceman from Buffalo to bring some added stability to the team. Warrener was twenty-seven, and had already played nearly 500 NHL games. "I vividly remember meeting Monty for the first time," Warrener says. On a team, names quickly become nicknames, and Steve, all his hockey life, had been "Monty."

"We had just done our first day of physicals at training camp. It was about six o'clock and I was in the restaurant of the Melrose Bar eating dry ribs. I remember that because I'd been living in the U.S. and you can't get dry ribs in the States. I was the only veteran there. The others were training camp guys trying to make the team. There were about six or seven of us. Monty and I were sitting next to each other. We seemed to laugh at the same jokes. Right away, we seemed to know what the other was thinking. We picked up on each other's moods. I remember at the end of the night thinking that this guy is going to be a friend. Then, I thought, it's too bad he's not going to make the team."

Undrafted out of junior, Steve had been signed as a free agent by the Flames in 2000. He played the 2000-01 season with their AHL team in Saint John, New Brunswick, before bouncing back and forth between Saint John and Calgary for the next two years. At the end of training camp, the Flames had to decide between Steve and Mike Commodore for their seventh and final defence spot. "Monty had played well, but I thought Commodore outplayed him," Warrener recalls. "But Jimmy Playfair had a soft spot in his heart for Monty, and the team kept him." Playfair, one of the Flames' assistant coaches, had been Steve's coach in Saint John three years earlier, when the team won the AHL championship.

The Flames began the regular season slowly. Roman Turek, their starting goalie, was injured in the second game and didn't play again until the new year. In mid-November, with the prospect of playoffs already slipping away, Sutter acquired goalie Miikka Kiprusoff from San Jose. Kiprusoff had seemed on the verge of stardom for several seasons but had never broken through. In December, with him in goal, the Flames lost only two of their next fourteen games.

Steve was playing even less than he had the year before—but surrounded by other young players and on a team beginning to win, he was having the time of his life.

"We had a lot of single guys," Warrener recalls. "Everyone hung out together." For them, all day—every day—was team time. Practice was at eleven at the Saddledome, or across the street at the Corral. It lasted an hour. Then to the gym, for twenty minutes if you were a regular and playing long minutes every game, as Warrener did; forty-five minutes to an hour if you were a healthy scratch, as Steve often was. The players all rode the stationary bike, then did some weights—then Steve hit the weights harder. Warrener tolerated the off-ice stuff; Steve loved it. In the gym, he felt as if he were a member of the team in a way in which, when the games began, he sometimes didn't. He knew that the training he did would somehow, some day, pay off for him and for the team.

"We had an attitude that year," Warrener recalls, "all of us, that 'When I go to the rink I am going to be the best player, whether it's in practice or training, and I'm going to work my tail off because, guess what, I love what I'm doing.'" And, he says, for "a young guy like Monty, you're not seen or heard, you just work, and [you] get in a game here and there and think to yourself, 'I'm here, and life's not too bad.'"

When practice was over, the team that skated and trained together went to lunch together. Typically it was Warrener and

Steve, and backup goalie Jamie McLennan; often they were joined by Regehr and Kiprusoff, sometimes winger Shean Donovan, and Commodore when he was up with the team—most of the unmarried guys. They went to the Mongolie Grill or Joey Tomato's. "It had to be a place that had chicken and white rice," Warrener laughs, because "Noodles," as McLennan's teammates called him, "was a very picky eater." Weirdness on a team is tolerated, even encouraged, because it can be mocked. McLennan was also a goalie.

Lunch was a time to talk about last night's adventures, and about the next game. Then it was home for a nap, then back together for dinner, then a movie. Two hours spent watching a film brought them two hours closer to the bar. And the stupider the movie, the stupider the lines they could riff on with each other for days after. *Dumb and Dumber* was a favourite from some years before, and *Caddyshack*, and that year's must-see was Jack Black's *School of Rock*. Steve and McLennan's particular obsession was *Kingpin*, the story of Roy (Woody Harrelson), a bowling prodigy fallen on hard times, now with a prosthetic bowling hand; Ernie (Bill Murray), an established, obnoxious pro-bowler; Ishmael (Randy Quaid), Roy's mentee, who is Amish; and Claudia (Vanessa Angel), the love interest. When Steve and McLennan would see each other, they would begin quoting lines from the movie. Either of them might start:

> ESPN Announcer: So, Roy, let me ask you, what have you been doing all these years?
> Roy: Well, I uh . . . Drinking. A lotta drinking.
> ESPN Announcer: Are you still drinking?
> Roy: No. I uh . . . Why, you buying?

And their favourite:

Ernie (in the direction of two girls together): Hi . . .

One Girl: Hello . . .

Ernie: Not you . . .

Ernie (to the other girl): Hi.

Years later, on different teams in different cities, "All of a sudden I'd get a text from Monty," McLennan recalls. "It was just 'Hi.' And I'd text back, 'Not you. Hi.'"

For Steve, McLennan, Warrener, and the rest, to do what they were doing three nights a week, eighty-two games a season, they needed to know each other, like each other, trust each other. What to non-players might seem like killing time between games, to them was hanging around with a purpose. The dressing room, ice surface, gym, restaurant, movie theatre, or bar was where they did the work of the team. It was their office. And sometimes, Warrener admits, they stayed too late at the office, "doing things maybe we shouldn't have." Girls, alcohol, even drugs for some . . . but they knew that the more they won, the more nobody—coaches, managers, fans, media—would see anything. For them, winning meant freedom.

"You're young and cocky, and you know everything," Warrener says, thinking back on that team. "You're a band of brothers."

"We had a great coach, Darryl Sutter," he adds. "He'd walk through the dressing room and right away he knew the mood of the team. Then he'd push the button that needed to be pushed." Sutter always talked about "chemistry," McLennan recalls. "He wanted players that he felt could fit into the room. He wanted to have good guys; guys that would buy in." And to Sutter, buying in meant a player not putting himself first. "To Darryl," as McLennan puts it, "if you didn't buy in, you didn't fit in."

McLennan remembers a rookie dinner early in the season in Chicago, when things got out of hand. "But Darryl just said to us what he always said: 'You deal with it.'" By that, of course, he meant deal with it by winning. "Because," Sutter told them at the time, "if you don't win, *I* will deal with it."

Warrener remembers another time when the players didn't deal with it. "We were playing Colorado, Kiprusoff was hurt, and Jamie was in net. We were down 2–0, it's the end of the first period, and Darryl came into the dressing room and starts yelling. 'You fuckin' guys. Your good buddy Jamie McLennan. You all care about him. He's the best. He's your friend. He's wonderful. He goes in the net and you lazy bastards won't play hard for him. Some friends you are.'" As Warrener recalls, "We all looked around the dressing room at one another and went out and won the game."

Steve played twenty-six games during the regular season. He would sit out for a few weeks, then play a game, then somebody would get hurt and he'd play four or five more games, then he'd be in the press box for ten days, then play another game. He didn't play more often because, as Warrener put it, "he wasn't as good as Regehr or Lydman or Gauthier or Leopold or Ference or myself. He was a skilled guy. He was a good player. But he was raw. He was just learning." And what Steve was learning was what he always seemed to be learning.

"Monty was confident," Warrener continues, "and that was a beautiful thing. But he was too confident. He'd be a healthy scratch for a couple of weeks, then go out and make an unbelievable play, then all of a sudden throw a backhand saucer pass through the middle of our zone and I'd be screaming to myself, 'Nooooooo, don't do it,' because if it doesn't work you're toast. You're back in the press box for a month." Sometimes Steve would remember what every coach had told him all of his life and not make that pass, and sometimes he'd forget. About once a game, he would have what McLennan calls a "hey, I'm Bobby Orr and I'm going to try something" moment.

To a coach, a good guy is someone who does what he's supposed to do. To a player, it's not much different. One guy makes a blunder; they all pay the price. But teammates also hope harder for good guys, and fear deeper. "Darryl knew how much we liked [Steve]," Warrener recalls. "The team loved Monty. He had his faults, for sure, but there was a lot more good about him than bad, and we could see that. Darryl really liked him, too. As hard-assed as he could be, and he was hard, Darryl could see who Monty was, and was pretty supportive of letting Monty be Monty. As long as it didn't cost us a hockey game."

"Monty was a good player," Warrener continues, "but really he was a better teammate. He made being on a team fun." A team needs good players, and it also needs good teammates—and, as a teammate, Steve didn't have lapses.

"I remember we were in Anaheim late in the season," Warrener says. "We had gone out for supper and were walking back to the hotel. There were ten or twelve of us. My mind was wandering and I started looking at these guys, and I said to myself, 'Craig Conroy, good guy. Marty Gélinas, good guy. Jarome Iginla, good guy. Miikka Kiprusoff, good guy. Shean Donovan, good guy. Monty, good guy. I just kept going down the line, and then it dawned on me, 'That's why we're having success. There's a bunch of good guys on this team.' We spent a lotta, lotta, lotta time together. We had Kipper and Jarome [Kiprusoff and Iginla, two star players] but we made our run because we cared about each other and worked our tails off."

The Flames finished sixth in the Western Conference and played third-place Vancouver in the first round of the playoffs. The teams split the first four games, each winning and losing a game at home. Lydman got a concussion in Game 3, and Commodore replaced him for Game 4. Gauthier tore up his knee in Game 6; Steve replaced him and played the seventh and deciding game of the series. In that game, Iginla scored two goals and the Flames went ahead, 2–1; the Canucks tied the score with only 5.7 seconds to go, then Gélinas scored in overtime to win it for the Flames. Steve played just over ten minutes, was plus-1 (he was on the ice, at even strength, for one more goal for than goal against), and never left the lineup for the rest of the playoffs. Two healthy scratches, Commodore and Montador (or "The Doors," as they came to be called)—one who had spent most of the season in the minors, the other in the press box, both of them putting in countless hours in their respective gyms preparing for this moment that they thought would never happen—were now needed as regulars.

When a team is on a roll, it doesn't even realize it's on a roll. It just rolls. For Steve, Warrener, Commodore, and the rest of the Flames, the playoff fun was only just beginning. And the more fun it was on the ice, the more fun it was off the ice. Flames T-shirts and hats were everywhere in the city. Inside the Saddledome was a sea (a "C") of red Calgary jerseys; outside, when the game ended and the fans poured out, a section of 17th Avenue became the "Red Mile." Oil patch mechanic, teacher, banker, or barista—in the playoffs, everyone becomes a player. And as the games went on, more and more green hard hats began to appear.

Craig Conroy had gotten the idea of a hard hat from Kelly Chase, his former teammate on the St. Louis Blues. The Flames were a team of two stars, Iginla and Kiprusoff; one all-rounder, Gélinas; and a bunch of grinders. The stars had to play like stars, and they did—but the team would only go as far as its grinders took it. Iginla and Kiprusoff had their own identities; the rest of the team needed their own. Conroy found a battered green hard hat abandoned in the catacombs of the arena. "We weren't a pretty team," he recalls. "That green hard hat symbolized what we were all about." The players awarded it after every win, to the "unsung hero" who had done the kind of grimy, scuffly stuff on the ice that those outside the dressing room might not have noticed. The player who had won it the previous game was the one who presented it to the next recipient, getting lots of advice from his fellow grinders. "We had a song and Rhett [Warrener] would do a little dance, and it was hilarious," Conroy recalls. "Then we'd turn off the music and give out the hard hat."

"Jarome and Kipper would *never* get it," Conroy says with pride. After all, they weren't grinders. But the presenter would often fake a toss in their direction, then throw the hat to someone else, and everyone would roar. The winner had to wear the hat for post-game interviews. Fans at home saw the interviews and began wearing their own hard hats to the games. This was the playoffs: the fans were grinders, too.

"Going into the season, we weren't supposed to do anything," Warrener says. "We got Kiprusoff, which saved us, then we beat Vancouver, so holy smokes, now we've actually done something. Everyone else in the world is saying, 'Hey, that's awesome, but now you're going into Detroit, who's got everybody, and now you're done.' But we knew we'd beat them."

The Wings had won the Stanley Cup two years earlier, in 2002—and with Nicklas Lidström, Steve Yzerman, Brett Hull, Brendan Shanahan, Pavel Datsyuk, and Henrik Zetterberg, had finished first overall during the regular season. The Detroit–Calgary series went six games and the Flames scored only eleven goals—including just two in the last two games, both of which they won, 1-0, the second in overtime (Gélinas again). In that final game, Steve played twenty-six minutes and was plus-1. More importantly, his teammates awarded him the Green Hard Hat.

Then, what couldn't get better got better. Calgary played San Jose in the Western Conference Final. The Flames went ahead 2–0 in Game 1, but the Sharks tied the score. Conroy gave Calgary the lead again, but San Jose scored with just over three minutes left, and the game went into overtime. Many minutes passed. The Sharks got possession of the puck in their own zone and started up the ice. Their forwards headed to the bench for a change. Suddenly, the puck got loose at centre and Iginla picked it up. He skated to the Sharks' blue line, it was two-on-two; behind them was a large gap as the San Jose players scrambled off the bench to get into the play. Steve saw the open ice. Iginla circled into the left corner; Steve rushed towards the net, banging his stick on the ice. "You could hear him in Sacramento," Sutter said later. Iginla, as *Hockey Night in Canada* announcer Chris Cuthbert, put it, "heard the beaver tail" slapping and passed the puck to Steve. He was all alone with Sharks goalie Evgeni Nabokov. Steve waited, and waited, and shot. The puck went off the post and in. The photo in the newspapers the next day showed Nabokov, his right leg split to the corner; Steve, his arms in the air, his mouth wide open (and minus his right front tooth) in a huge, over-the-moon grin. In the next moment, Steve turned and ran on his skates up the ice, his teammates running after him. They caught up to him, and together they piled against the boards.

"It makes me emotional just talking about it," Warrener says now. "Because we all loved him, and because he worked so hard that year. For him to do that, and in that Monty way, banging his stick coming down the ice. It was perfect. And that look on his face when the puck went in. It was just awesome." Warrener pauses in his recollection. "For me, that play epitomizes his life. It took confidence. It was overtime. In the Stanley Cup playoffs. He saw a small window to jump up, and he did."

"To me the goal is one thing," Gélinas says, "but the smile is another. He was just so happy. Everything was just so special. I will remember that moment for the rest of my life."

The Flames won the second game in San Jose, then lost both games in Calgary, then won the next two to end the series. The Red Mile got longer and longer. The Flames had beaten all three Western Conference division champions. Gélinas had scored all three series-winning goals, two of them in overtime, and became known as "The Eliminator." But Warrener has another memory of the events. "We won both games in San Jose, then blew two at home. To play extra games when you don't have to really hurt us. We should have done a better job of putting them away."

Tampa Bay was next, in the Stanley Cup Final. The series went seven games, and at one point the Flames were ahead three games to two. They could have won the Cup—but they didn't. The players all live with memories of what they could have done differently. Warrener and Gélinas remember "the run in 2004" with regret and with pride. And they remember the fun.

McLennan watched the playoffs on TV. He had been traded late in the season to the Rangers. He remembers Steve in those games: "Monty was the type of guy who embraced everything, so he embraced the whole package of it. He'd say to himself, 'I'm not just going to enjoy the games, I'm going to enjoy the fanfare, I'm going to enjoy the popularity, I'm going to enjoy *all* of this.'"

CHAPTER THREE

How did we get to where we are? To hockey as we play it, to Steve, to brain injuries and CTE?

The story starts at McGill University in Montreal, in 1875. For centuries, people had been strapping bones or wood or metal objects onto their feet to move across the ice. But on March 3, 1875, eighteen McGill students, many of them rugby players, came onto the ice at the Victoria Skating Rink near the university, in order to play a very special game. It had been some time in the planning, and an announcement had appeared in the *Gazette* that morning.

Similar games had been played on outdoor ponds for decades in Canada, most notably in Nova Scotia. The sticks of the early players were short, like those in field hockey; a lacrosse ball was used in lieu of a puck. The players acted as their own organizers: the rules evolved by trial and error to reflect the game they wanted to play and the conditions they faced. Depending on the size of the pond, any number of players could play.

It was different indoors on the Victoria Rink. The ice surface was smaller—there was room for only nine players on each side. There were no boards; only a slight rise separated the ice from the off-ice area around it. To protect the rink's glass windows and the large number of spectators expected, the usual lacrosse ball was considered too dangerous and a flat, square, wooden "puck" was used instead. The "nets" were two vertical sticks with flags on top, set eight feet apart. The game was played in two thirty-minute halves. As in rugby, player substitutions were not allowed to enter and re-enter the game, and all of the players remained on the ice throughout the entire sixty minutes of play. After all, sport was a test of endurance. Also as in rugby, no forward passes were permitted. The puck could be advanced only by a player skating it up the ice. After all, sport was a trial of character.

The players' skates offered little support, and their blades were clunky; in fact, these first hockey players skated like rugby players. But they moved faster on skates than anyone could move in shoes, so the game seemed fast. They collided with each other and fell to the ice often, so the game seemed dangerous. They wore no other equipment. After all, sport was a test of courage. To a remarkable extent, because of circumstance—the skates; the ice surface that was 204 feet by 80 feet, almost exactly the dimensions of today's NHL rinks—and because of happenstance—the speed and the collisions between the players—and because of the sporting understandings of the time— no forward passing or free substitutions being allowed—the future game of hockey was on display that night in the Victoria Rink.

Two years later, in 1877, the "McGill Rules" were codified, and published in the *Gazette*. The future risks of the game were antici- pated: "No player shall raise his stick above his shoulder. Charging from behind, tripping, collaring [holding], kicking or shinning [slashing] shall not be allowed." Years of informal, evolving play had shown that a stick might be a weapon, that a stick or a body might send an opponent dangerously out of control, that the head needed to be protected.

In 1893, Lord Stanley of Preston, Canada's Governor General, gave as a gift a trophy to be awarded annually to the amateur hockey champion of Canada. The Dominion Hockey Challenge Trophy came to be known as the Stanley Cup, and would have an effect far beyond what was originally imagined and even intended. Players suddenly had something to focus on and aspire to. More people began to play, and the game spread into more and different parts of the country. New indoor arenas were built, which meant more games could be played as fewer were weathered out. By 1910, games were divided into three periods, rather than halves, allowing the rink to be resurfaced twice during each match. The ice got better. Skates got better. The players and their skating improved. If the puck could be pushed up the ice, why could it not be stickhandled? If collisions happened, why could hip checks not be practised and orchestrated? Hockey became more challenging. In 1911, it changed from a seven-against-seven game to six-against-six; free substitutions were allowed. The players, now on the ice for shorter stretches, could skate harder. The game got faster. With better players and a more exciting game, spectators were willing to pay to watch, and so professional teams were created. Players could afford to play more hours in a week, more years in a career—and they got better still.

In 1917, the NHL was formed. All of its founding four teams were in Canada, but hockey was popular in prep schools in the U.S. northeast, too. After all, if the future could be won on the "playing fields of Eton," as the British elites liked to believe, why not in harsh winter conditions on the ice rinks of Exeter and Groton? So the

NHL expanded into these hockey regions, with teams in Boston, New York, and Pittsburgh.

As the game grew off the ice and speeded up on it, its pace of play also began to slow down. If an advancing player was allowed to pass only backwards, opponents could retreat and congest the defensive zone. Space and time began to disappear; goals became harder to score. During the 1928–29 season, goalie George Hainsworth of the Montreal Canadiens recorded twenty-two shutouts in forty-four games. On average that year, opposing teams *together* scored fewer than three goals a game. Professional teams needed fans; fans wanted goals. During the 1929–30 season, the forward pass was allowed. Within a few years, scoring had almost doubled. The game got even faster.

Yet it wasn't as fast as it might have been. Players accustomed to playing without substitutions found it hard to come off the ice after only ten minutes or so. Players used to stickhandling up the ice still passed only when no other option was available. And a pass, of course, couldn't be allowed everywhere. One from a player's own defensive zone to a teammate near an opponent's net was contrary to the spirit of sport. The puck, after all, couldn't be allowed to do all the work. So blue lines had been introduced, repositioned closer or further from each goal-line (the red line wasn't added until the 1940s), which limited where passes could be made and forced players to skate more slowly than they could. The speed of the game couldn't be entirely uncontrolled. After all, sport required discipline.

In the 1930s, the players were still mostly of average size for a North American adult male. The greatest star of the decade, Howie Morenz of the Montreal Canadiens, was five-foot-nine. Charlie Conacher and Red Horner of the Toronto Maple Leafs were giants, at just over six foot; neither of them weighed 200 pounds. Almost no players wore helmets—those who did had usually suffered head injuries but wanted to continue to play. Injuries were primarily facial cuts and lost teeth; cuts healed, and missing teeth were hardly unusual for adults at the time. But the consequences of more severe injuries were much greater than today. Medical care was more basic; medications were less advanced. Several players lost eyes. Ace Bailey of the Leafs had his career ended by a head injury. Morenz died from complications of a broken leg suffered in a game. Countless others had their careers shortened or diminished by ligaments that were never repaired, or never healed. But, in general, this was an age of risk. People worked on farms, in factories, and in mines where life-threatening accidents were a regular occurrence. Hockey injuries were just part of the game, and part of the era.

The game's rules did offer some protection. A hit from a hip or a shoulder was considered fair. Both were blunt instruments, and a hip could strike an opponent's head only if the opponent's head were down. In such cases—the thinking went—the player deserved it. Stickhandling, a player's principal offensive weapon, could be done better if he looked down at the puck. If he sought this as his advantage, he should have to live with its consequences. Besides, in this much slower game, forceful collisions were infrequent.

Elbowing and kneeing penalties were introduced, as elbows and knees are pointed instruments and, when thrust out, allow a player to extend his body and strike an opponent at his two greatest vulnerabilities—his head and his knees. Charging (from the front; the McGill rules had covered charging from behind) and boarding were also penalties. Taking a few running steps to make an offensive or a defensive play was fine, but doing so to hit an opponent, or hitting an opponent a short distance from the boards and throwing him uncontrollably into them, was considered too dangerous.

A player's purpose in both cases could be only to hurt—if not to injure—his opponent.

Helmets didn't become mandatory for new players entering the NHL until 1979, the year Steve was born. For the most part, players protected themselves by the way in which they played. They skated with their heads up, partly in order to see everything, but also so that their heads were further from the ice and less in range of sticks, shoulders, and elbows.

In 1959, Jacques Plante became the first goalie regularly to wear a mask. Goalies had always employed what was called the "stand-up" style. Initially they had been penalized for lying on the ice—it being considered unsporting to stop a puck merely because shooters, with their heavy, inflexible sticks, couldn't raise it high enough to get it over them. Goalies who sprawled on the ice were later disparaged as "floppers." The stand-up style was understood as the proper, most effective, and only way to play. But it was also the safest way. In stand-up position, a goalie's head was above the top bar of the net, and so less in the line of dangerous fire.

Players found little protection in the equipment that was allowed by the NHL, or anywhere. In time, skaters could put small, form-fitted cups over their shoulders, elbows, knees, and groins, and goalies could wear leg pads like those a cricket wicketkeeper used, but neither could wear padding that took away much more than the sting of pain. A body check hurt the hitter almost as much as the player being hit. Collisions with opponents or with the boards affected a player more the faster he went, and moving faster left him less time to see what was coming—less opportunity to avoid the blows—resulting in even more dangerous collisions. Speed had consequences.

Speed was hockey's greatest risk. Yet hockey's greatest appeal was speed, and the excitement it generated. The question was how to keep the two of them—the excitement and the risk—in balance. Players were allowed to slow down the game on their own. They could "rag the puck," or "freeze" it against the boards; neither the officials nor their opponents would put much pressure on them to keep it moving. The goalies could hold onto the puck when they chose, too. Mostly, however, it was the practicalities and customs of the game that slowed the player down—the quality of his skates and of the ice, the long shifts, the blue lines, and the game that generally moved at the speed of the puck-carrier rather than the puck. Hockey may have seemed an all-out, "damn the consequences" game, but it was always a compromise between performance and safety.

This was also true even in its fighting, which had always been an element of the game. In football, baseball, and basketball, a fight brought its combatants an ejection from the game; in hockey, the punishment was only a major penalty. The theory was that in hockey, with players moving at high speeds in an enclosed area, collisions were inevitable—and some of them, whether by accident or on purpose, would be judged unfair by the offended player, generating anger and the desire to get even. A player's alternatives were to use his stick against his helmetless opponent, or his fists. As stick-swinging was considered much more dangerous, fists were allowed. But as fighting was permitted only because of the "righteous" anger it provoked, the offended player had to fight his own battle. And few players fought very often—fighting on skates being difficult for anyone. As a result, most fighters were not very adept— little damage was inflicted by either party—and penalties for fighting were for five minutes, not more.

But because hockey was faster and more dangerous than other games, hockey players developed a reputation for their toughness, one that they wore with pride. They were expected to play through everything, and were known for putting "everything" into a game. But what is considered *everything* is different depending on the era. In the mid-1930s, it was a forty-eight-game regular-season schedule, and three rounds of significantly shorter (best-of-three, best-of-five) playoff series. It was seasons that began in November and ended, *after* the playoffs, in early April. It was less intense practices and no off-ice training, and five-month off-seasons with little or no training at all. Players who had grown up on farms or in working-class homes—Morenz, Maurice Richard, Gordie Howe—had the advantage. Their off-ice, off-season training consisted of doing chores that involved much more than just taking out the garbage. Because players earned a normal worker's wage, *everything* meant working another job in the summer. It was also retiring from hockey and having another career for the rest of their life.

Until the early 1960s, *everything* didn't change much. The seasons had become longer—seventy games—but in a six-team league, the playoffs were only two rounds. There was still no off-ice, off-season training for the players. Players now earned more than an average worker's wage and many no longer had summer jobs, but all of them would need second careers, usually outside hockey, to carry them through the last thirty-plus years of their working life. The medical care they received was better, but not much better; injuries were still mostly cuts and lost teeth. The game's penalties continued to focus on the head and the knees. Fighting remained between the aggrieved and the "aggriever"— two combatants little capable of inflicting damage. Equipment offered more (and better) padding, but still took away little of the pain. On average the players remained of normal North American adult-male size—Elmer "Moose" Vasko, who retired in 1970, was six-foot-two and 200 pounds. (Too tall to be in the range of an opponent's high stick, Vasko played his entire career without ever losing any teeth.) Farm and working-class kids with their off-ice, off-season hard labour of chores retained a natural advantage.

On the ice, the game had gotten faster, more exciting, and more dangerous—but not by much. Coaches and players still hadn't realized most of the possibilities of free substitutions and the forward pass. The puck carrier was still king; the (blue, and now red) lines still held back play; shifts grew shorter but remained two minutes long even through the 1950s. In highlight clips and in memories, Richard, Howe, and others skated a hundred miles an hour. In full-game films, the players—Richard and Howe included—coasted a lot, burst forward or back when opportunity arose or when need be, and mostly paced themselves. The action was very slow. Other than in the scoring areas near the net, the players had lots of time and space to create plays and protect themselves. The game was still a compromise between performance and safety.

Three events in the 1970s changed the game: the Summit Series in 1972; Stanley Cup wins by the Philadelphia Flyers in 1974 and 1975; and the creation of the WHA in 1972 (and its subsequent demise and the absorption of its four remaining teams into the NHL in 1979).

The Summit Series of 1972 pitted for the first time the best of the so-called amateurs—the perennial world and Olympic champions from the Soviet Union—against the best of the professional world, the top Canadian players in the NHL (and the top players in the NHL *were* all Canadian). The series began on September 2 in Montreal, with a stunning 7–3 loss by Team Canada in Game 1. It ended twenty-six days later in Moscow with Team Canada's almost-as-stunning 6–5 victory in Game 8 to take the series. Yet a bigger surprise was ahead.

The Soviets had begun playing organized hockey only twenty-six years earlier, in 1946. In order to someday challenge the superiority of Canada, Soviet players knew they had to work harder and longer. They needed to accept their own sporting traditions and the conditions of Soviet life—just as Canadian players had done in Canada—and make the most of both of them. The Soviets had little indoor ice, and so they had to develop their skills mostly through off-ice training. This they could do just as well in summer as in winter; they could train eleven months a year.

The traditional winter stick-and-ball sport in the Soviet Union is called bandy, which is similar to field hockey on ice. It was played on a soccer-sized pitch too big for players to skate a bandy ball from end to end themselves, so passing has to be the focus. The Soviet Union's first hockey players came from bandy, so in Soviet hockey the pass receiver, not the puck carrier, was king. In the Summit Series, the Soviets did not win, but they showed that hockey could be played at the highest level in more than one way.

This was their crucial achievement. Out of the necessity of their own circumstances—and the traditions of their own winter sport—they demonstrated the possibilities of the forward pass, and of off-ice, off-season training. And so the Summit Series produced an almost unique result: its winner was changed more than its loser. While the influence on Canadian hockey wasn't visible for almost a decade, it shaped the hockey lives of all the players that came after—including Steve Montador's.

The Philadelphia Flyers had entered the NHL as one of six new expansion teams when the league doubled in size in 1967, but the Flyers were disadvantaged in a way even the other expansion teams were not. The existing "Original Six" teams made very little available to the new franchises in the expansion draft in terms of skill or promise, but while the other new markets (Minnesota, Pittsburgh, Los Angeles, California/Oakland, and St. Louis) had some history of hockey success—at least at the box office—Philadelphia had none. The Flyers would need to compete on the ice with a team that wasn't competitive, and compete off the ice for new fans. It was very possible that some of the new teams, especially Philadelphia, would fail.

In 1971, the Flyers hired a smart, cynical, and pragmatic coach, Fred Shero. To win and to give the Flyers a chance to survive, Shero had a choice. He could try to make the team a lot better fast, which he couldn't do, or he could try to make his opponents worse.

Intimidation had been a strategy used throughout hockey's history—by individual players, and by some teams against certain opponents. But for the Flyers, intimidation became the basic team approach. A superior player who is not focused completely on what makes him superior, but is instead distracted by concerns for his safety, is no longer superior. The Flyers brought their opponents down to their level, then beat them with their own sprinkling of stars: inspirational captain Bobby Clarke, goalie Bernie Parent, and scorers Bill Barber, Reggie Leach, and Rick MacLeish. The "Broad Street Bullies" were born.

Other teams had been tough; the Flyers were punishing. Some players, as healthy as butcher's dogs until they noticed the Flyers were next on their schedule, suddenly came down with the "Philadelphia flu," as it came to be known, and were unable to play. For the Flyers, violence became *the* product, not a by-product, of their game. Fighting wasn't about one player righting a wrong, but a team event. The Flyers won the Cup in two straight years, 1974 and 1975, and because losers imitate champions, the league got tougher, fighting increased dramatically, and players got bigger.

The game also got faster. Expansion-team players were less skilled but could skate as fast as Original Six players—in straight lines, especially if they didn't have the puck. So the strategy for expansion-era players was to skate to centre ice with the puck, dump it ahead into their opponent's corners, and chase after it. It was called "forechecking." The result was less coasting, more bursting; more frequent collisions of greater force; more tired players, and, as a consequence, shifts that weren't much more than a minute long.

While the Flyers' bully-effect diminished in the years ahead, it never entirely went away. The definition of putting *everything* into a game was changing.

In 1960, the American Football League (AFL) was formed, allowing more cities to have a "major league" team and to compete with the National Football League (NFL). The American Basketball Association (ABA) was created in 1967 for the same reasons. Five years later, in 1972, the World Hockey Association (WHA) began play. All of these rival leagues disappeared within a decade of their founding, but some of their teams were absorbed into the original leagues, and each would leave its mark.

They all began needing to get attention, so they took chances: the ABA introduced the three-point shot; AFL coaches brought in new offensive strategies; and the WHA signed several European players. For the first time, NFL, NBA, and NHL teams had to compete for a player's services. This was a development that proved hugely significant: players had a choice, and salaries exploded. In hockey, they more than tripled within a decade. And more money and more teams meant more dreams for more players—and more opportunities to realize them. Hockey, which had changed only little by little in its first one hundred years, was about to be transformed. More money meant higher stakes for players; more off-ice, off-season training meant stronger, better-conditioned athletes. For the next generation of players, hockey would be even faster and more exciting. It would also be more dangerous.

Less than three months after the newly merged league began its season, and Wayne Gretzky played his first NHL game, Steve Montador was born.

CHAPTER FOUR

Steven Richard Montador was born on December 21, 1979 in Vancouver General Hospital. He weighed seven pounds exactly, and had brown hair and blue eyes. His father, Paul, was with his mother, Donna, in the delivery room. His three-year-old brother Chris and his sister Lindsay, seventeen months—along with Paul's parents, who lived in Vancouver, and Donna's, who lived in Edmonton—all visited him the next day. In Steve's baby book, under "News headlines of the day" for December 21, Donna recorded, "Simmer snaps record. Flyers tie one." A few days later, the birth announcement to family and friends read: "The forward line is in place. Now we have to work on the defence."

The Montadors lived in Richmond, a fast-growing suburb on the Fraser River delta, about fifteen kilometres south of downtown Vancouver. Paul was thirty-three and a rising young executive in the pharmaceutical industry; Donna, at thirty-two, with three kids age three and under, had her hands full at home.

Donna had grown up in Edmonton. Her grandparents had left

the turmoil of Eastern Europe for the unknown of Canada not long after Alberta became a province in 1905, settling on a parcel of wild grasses and trees near Smoky Lake, about 100 kilometres northeast of the city. When Donna's father was old enough, he moved off the farm—first to a series of nearby small towns, and finally to Edmonton to sell what a post–Second World War public was ready to buy: insurance and real estate. He also loved hockey. When the WHA was formed and the Oilers came to town, he was "second in line" to buy season tickets, Donna remembers. He bought four of them, just above the home bench. Donna's friend, Barb, dated Al Hamilton, one of the Oilers' players. In the off-season, Hamilton played fastball. Paul Montador, newly arrived in Edmonton from Vancouver, loved baseball and—not knowing anyone outside of work—joined Hamilton's team. Through Al, he met Barb; and through Barb, he met Donna. Al Hamilton, a decade later, would be the first Oiler to have his number retired.

Paul had come to Edmonton to work for Ortho, a subsidiary of Johnson & Johnson. It was his job to travel through central Alberta, Yukon, and the Northwest Territories as a "detail rep," selling the company's birth control products, calling on doctors, and setting the table for the sales reps who would later close the deal. It was his first job out of university.

He hadn't wanted to leave Vancouver. All his family was there, including his grandfather, with whom he was close. His father had been the head of building approvals and inspections for the City of Vancouver, at a time when City Hall was an important place to work. As the family story goes, he was also a local baseball star who had once pitched a twenty-three-inning shutout against a barnstorming U.S. team, winning 1–0. His opposing pitcher: Satchel Paige.

Even more so, Paul didn't want to leave the city because Vancouver was Vancouver: the mountains, the water, the weather. To him, there was no place like it. But at Ortho he would get good training. He would learn how to sell—a necessary skill in any job, he believed. He would get to know an important industry and not have anyone perched on his shoulder telling him what to do. Nor did he want to live the office-bound life of his father. Besides, Edmonton wasn't forever. Paul had big things in mind for his life. Nothing specific, but big.

Paul loved to learn, and wanted to be in positions and places where he *could* learn. He wanted to know more about whatever he was doing than anyone else around him. And he wanted to be in charge. He had learned that in baseball. He was a catcher, and as a catcher he was involved in every play; he set the tone; he had to think ahead. His job was to keep everyone in the game. If his pitcher was drifting, he would fire the ball back to him hard enough to sting his hand and get his attention. He had to be the manager on the field. It wasn't long before he realized that he liked managing as much as he did playing. In whatever job he took, in whatever field, he was going to live a manager's life. Ortho was the first step, even if that meant Edmonton.

After high school, Donna had taken some business courses, then got a job with a travel agent. Paul was on the road a lot during the week with his work, and on weekends he played in fastball tournaments around the province with the Payton Playboys, Alberta champions for two straight years. His new relationship with Donna seemed to be going somewhere, then it didn't, then they didn't know. When it reached one stuttering point, Donna heard that Wardair, a charter airline, was looking to open up routes across Canada and to the U.S. and Europe, and they needed a sales agent in Hawaii. Donna was offered the job.

Paul, too, was getting restless. At that point, he felt that he had learned all he was going to learn at Ortho and was growing tired of Edmonton. So when he was offered a job in Vancouver as a sales rep at Canlab, a large hospital supply company—effectively one step higher up the ladder—he took it. He and Donna went their separate ways, but absence soon had its effect. Over Christmas in 1971, Paul flew to Hawaii and he and Donna became engaged. Five months later, they were married in Vancouver. A year later, they moved to Toronto. Paul had been promoted again.

Paul believed that he learned about 80 to 90 per cent of what he could learn in any job in two years. Then it was time to move on: within the same company, if possible; or to somewhere else, if that was better. In Toronto he was an operations manager at Canlab, involved in purchasing and customer service—"a complete shift from sales," as he puts it. For Paul, it was another "learning opportunity," and a chance to build a broader foundation for his career. Donna was working at Wardair's Toronto office. Two years later, Paul was promoted to sales manager, then again two years after that, as the company moved more broadly into the hospital sector. It was 1976, the year Chris, their first child, was born.

Soon Paul was asked to run Canlab's new hospital division. There he worked with hospital architects who needed to know more about new medical technologies and their uses, just as he needed to learn more from the architects about the requirements of hospital design. His work took him across the country. Paul and Donna's second child, Lindsay, was born.

These were the prime learning years, as Paul believes—from age twenty-seven to thirty-two. Years to gain experience, to develop skills and confidence, to work new jobs, to live in new places. Within eighteen months, he was asked by Canlab to run hospital supply for all of Western Canada. It was a chance for Paul to move back to his beloved Vancouver. The family bought a modern, suburban home in Richmond; Paul bought a new white Corvette. A few months later, Steve was born. Less than a year after that, Paul returned to Toronto to run Canlab's hospital supply division for all of Canada. He was now working at the company's head office, in the country's biggest city; at Canlab, he had no place else to go. Now thirty-four, it was time for Paul to settle down.

They bought a house on Avonbridge Drive in Mississauga—Paul, Donna, Chris, Lindsay, and Steve—south of the Queen Elizabeth Way (QEW), the main east–west highway between Toronto and Buffalo, and just north of Port Credit, on Lake Ontario. It was in an area canopied by old maples, beeches, and pines, its windy streets shaped by the nearby Credit River and by Avonbridge Creek, which ran behind their property. Their street wasn't much more than a hundred metres long. It had new houses on both sides, and a big round cul de sac—"the circle" as they called it—at the end, where the Montadors lived. The circle, about thirty metres in diameter, flat and smoothly paved, was the gathering place for the neighbourhood's kids, for ball hockey games, soccer games, bike riding, skipping, hopscotch, mini-car racing; it was where games of hide-and-seek and capture the flag began. It was for boys and girls of all ages, the older ones amiably ignoring the younger ones who worshipped them; a place where kids played, and parents didn't worry.

Their house was only a few years old when they moved in. It was Cape Cod in style, with red brick, a porch, white trim and dormers across the roof in the front, and a shade-filled yard with trees and bushes and a large, curvilinear pool with a diving board in the back. From the pool, the yard sloped further downward—a perfect first hill for tobogganing—to a narrow, shallow, slow-running creek—a perfect first skating rink in winter. It was on this creek that Chris, Lindsay, and Steve learned to skate, though as Donna recalls, Steve spent most of his time lying on his back, looking up at the sky, chipping away at the ice with the backs of his skate blades.

Inside, the house was spacious and open, just right for a young family who put more priority on playing than on the scratches and dents that sometimes resulted. Steve's baby book says that he took his first steps in this house on December 19, 1980, two days before his first birthday, and that three days later he began climbing the stairs that led up to the second floor. There is a picture of him in the family photo album on those steps a few months after that, with his first black eye. He was round and chubby as a baby; his eyes, slightly close together like Will Ferrell's, were sometimes almost slit-shut in a grin, and at other times were wide open with a pensive, distant look.

The Montadors were a prosperous young family on the move. Chris was a big physical kid; he was all boy. Lindsay, almost two years younger, had to fight to keep from getting run over or ignored. Steve, three years younger than Chris, wanted to do everything his brother did, and he tried. The three of them were always doing something—and in their house, in their neighbourhood, there was always something to do.

In his first hockey photo, Steve is not yet three. It shows him in the basement family room, an oversized hockey helmet on his head, his face almost invisible behind its protective cage. In a photo taken a few months later, again in the family room, he has on old goalie pads that extend almost as high as his armpits; his blocker is turned backwards and his catching glove is twisted, as if he has been propped in place by Chris and Lindsay so that they can blast shots at him.

But in the photographs that follow, Steve's look soon changes. He moves out of family-room hockey to the backyard creek, and then to real-life arenas. He puts on full equipment, wearing it as if it were his own, with an excited, "can you believe I'm out here doing this?" expression on his face. He becomes old enough to join teams— St. Lawrence Starch, Ontario Gypsum, Dominion Securities, Royal Canadian Legion, Mooredale Construction. In team photos, he is usually standing in the back row, but never in the middle, and only once with a *C* on his jersey. In an individual photo taken when he was seven, a right-handed shot, he is positioned a little bent forward, his stick on the ice as if he were told to stand this way. A year later, crouched forward even more, a big smile on his face, he looks as he would in promotional photos years later, in junior hockey with St. Mike's, and in the NHL. He looks like a hockey player.

Within a few years, his face gets thinner and more mature, his shoulders become square and broad. At eleven, he has a high brush cut and a ready-to-please grin; at twelve, his hair is long enough to be slicked back, his hands are in his pockets, and he has a teenager's cool "okay let's get this over with" look on his face. There is another photo of him that year, taken at the bottom of the stairs that led to the second floor. Donna is next to him, standing on the first step; Steve is taller than his mother.

Growing up in a nice house in a nice neighbourhood—with his parents, his brother and sister, and his friends around him—it was a kid's life until he was twelve.

Then it became a hockey life.

For many years, Paul would arrive at the office every Monday morning to stories from co-workers, each one more horrible than the next, about their lives as parents of sons in AAA hockey. Stories about the games that might be played anywhere in the sprawl of Greater Toronto, and journeys in traffic that was getting worse by the month. About the weeknight practices that were too early for parents to put in a full day's work, too early for their kids to have a good dinner, too late for their kids to have a good night's sleep. About the Thanksgivings, Christmases, and school breaks when the games went on uninterrupted, interrupting everything else; about the weekend tournaments a hundred kilometres or more away in who knows where, the crummy motels, the coffee, the Timbits. About the injuries that their kids wore like badges of honour: bruised shoulders that they called "separations," twists that were "slight tears," and bumps to the head that they shrugged off as nothing because that's what the NHL guys did. There were stories about real or imagined traumas—the linemates who never passed the puck to their kid; the coach who didn't like him. The cost. The *time*.

One Monday morning, as Chris was nearing that hockey-committed age, Paul declared to a co-worker, "Over my dead body will my boys ever be involved in this insane activity."

This was a declaration Paul could make at work, and attempt to manage at home. He decided that instead of saying "no" to hockey, he would get his kids to say "yes" to something else. He was from B.C.—why not skiing? It was outdoors. It was healthy and active; and it was only on weekends and holidays, and only as long as the snow stayed frozen on the ground. It also meant trips to beautiful places. The family together. It was something they could all do all of their lives. It made sense. But Paul soon realized that skiing wouldn't be his answer. Really, there was no answer. This was Canada—neighbourhood kids playing ball hockey in the circle; NHL games on TV; team jerseys, toques, and T-shirts for Christmas and birthdays. Hockey was everywhere.

At aged twelve—"against my better judgment," as Paul recalls—Chris joined a AA team. It still wasn't that much more than it had been in house league and select, a game or two a week, a practice, a few tournaments; a schedule that started late enough in the year and ended early enough to allow time for soccer or baseball or trips with the family or messing around with friends. Chris was good enough to play AAA, but that would mean too much of everything—games, practices, *time*—for him and for the whole family. If Paul couldn't keep Chris from playing, at least he could manage the worst of the fallout.

But then, Chris got a chance to play AAA bantam with the Toronto Marlboros (known as the Marlies), the most storied minor hockey organization in the country. Its big team, the Junior A Marlies, had won the Memorial Cup as Canada's national champions seven times, more than any other team. Over two hundred Marlies players had made the NHL, including Hall of Famers Charlie Conacher, Joe Primeau, George Armstrong, Bob Pulford, Brad Park, and Mark Howe. The Marlies, as Paul put it, were a "great brand."

Paul decided that he needed to get to know the team's coach, Jim Nicoletti. If Chris were to sign with the Marlies, he would spend countless hours with Nicoletti—intense, meaningful, unforgettable hours. Nicoletti would become an important adult in Chris's life. The decisions that he made, the tricky moments he faced and the way in which he faced them, Chris would see. Chris would learn from him; but would he learn what he *should* learn?

Nicoletti was a school principal as well as a coach, and a family man with "real integrity," as Paul described him; his son, Dan, was on the team—a good kid and, unlike many coaches' sons, a good player as well. Besides, Paul started to think, even if Chris playing AAA didn't seem quite right for the family, to play at a level less than he had the ability to play didn't seem right either. After all, that's not how Paul had lived his life. And so Chris joined the Marlies.

Paul and Donna had a rule in the family. No matter what happened the rest of the week, no matter how much Paul had been on the road and how busy the five of them had been with their own lives, on Sunday they had dinner together. With AAA hockey, that rule went out the window. It was now a hockey life for Chris—and a hockey life for the family.

In AAA hockey, there are more games—two or three a week, plus weekend tournaments—and more practices, taking place in more months of the year, in more distant parts of the city or region. Chris's teammates now lived many kilometres away, in Scarborough or North York, rather than down the street. There was no carpooling to a game or practice. Nor could they take public transit. And because Chris had to be there for every game or practice—in AAA there are no excuses—Paul and/or Donna (with Paul's travel schedule, it was more likely Donna) had to be there—so Lindsay and Steve had to be there, too. Lindsay remembers a writing assignment that she did in Grade 4. She called it "My life in a hockey arena." Chris's commitment had become the family's commitment.

Then Steve got to be old enough to join a team, and he too moved up, from house league to select. Paul and Donna put him into power skating lessons with a woman who was a former figure skater. She taught Steve and the other kids in her class about inside and outside edges. She pushed them to try things they had never done before. She demonstrated to them a turn in which they would need to angle their bodies closer and closer to the ice, uncomfortably close, until it seemed as if they were almost sideways to it, with nothing to hold them up. The kids had to be willing to accept this feeling, and learn to deal with it. The instructor watched them as their bodies fought their minds—their minds winning, their bodies jerking upright—and finally said to them, "You've got to be willing to fall to learn how to do this turn. *You will have to fall.* There's no other way." Only Steve was willing to fall. That's how he was in everything, Paul recalls.

When Steve was twelve, he joined the AA Mississauga Braves. His coach was Jim Donaldson: "Mr. D," as the players called him. Donaldson had been a good minor hockey player in Toronto as a kid—an opponent or teammate of many who went on to have NHL careers. He was tough, and a good skater with a hard shot, but more than anything, he loved to play. He loved the trash talk and the laughter of the dressing room, the sweat-and-grin of practice, the give-it-all feeling of the big game. He loved being around the guys. He had dreamed of making the NHL himself, of course, but when he didn't, his hockey life merely shifted direction. He became a father and a coach. He coached mostly the way he played, but now with a parent's perspective—he and his wife had five kids, and his son Dave was also on the Braves. He knew that his players had their own dreams of making the big time, but his dream for them was to experience the full hockey life—the sweating, grinning, trash-talking, laughter, and give-it-all feeling—the one, years later, that he has come to value the most.

Donaldson turned forty-eight the year he coached Steve. He was working in the warehouse at Chrysler at the time, an eighteen-year veteran of the night shift—10:30 p.m. to 7:00 a.m. He had turned down promotions that would let him work other hours because, as he explains, "The night shift was perfect for me. It allowed me to coach." During breaks at work, he drew up drills for the next day's practice on IBM cards that were always lying around. On nights when his team wasn't playing, he sharpened skates at one of two local arenas, Huron Park or Clarkson. He did that for twenty years. Later, when his kids grew up and he stopped coaching, he moved to the day shift, but found he couldn't sleep.

Donaldson coached with a growl in his voice and a tight smile on his face, and when he spoke, his mouth turned slightly to one side, like tough guys used to talk. Most of the kids on the team had moved from team to team with him, from select to A, and now to AA. Nine of them would stay together for six years, which in a big city is something almost inconceivable in minor hockey. Steve was one of two players who were new to the team. He was also the youngest. With his December 21 birthday, a teammate or an opponent might be almost a year older than he was—not an insignificant advantage when you are only twelve years old. That older player can be more physically, emotionally, and intellectually developed. He is able to do more things, play with better players as linemates, be on better teams with better coaches. He is more likely to seem to others—teammates, coaches, parents—and to himself, to be a better player, and he is more likely to *be* a better player, both now and in the future. The first stars, usually the oldest kids, are often the future stars.

Steve had to find a way to fit in with his new team on the ice and off. But if anyone was worried he wouldn't manage it, they needn't have. Donaldson describes how, after a week, "I walked into the dressing room and there he was in the middle of the fray. He wasn't exactly shy and retiring." When Donaldson thinks about Steve later in that first season, he remembers him as "like [golfer] Lee Trevino walking up the fairway to his ball—talking to the fans, joking. But when he got to that ball, he would focus. It was his time. In practice, Steve would be kibitzing around, but then when I'd blow my whistle he was all business." He did every drill to "his absolute maximum. Good mood, bad mood, when he came to the rink, no mood. Just hockey. This was like heaven for him. It was where he wanted to be."

Steve had to pick up the pace of his game; this was AA now. He was a good skater, and had a good shot, but just because he did— and though he was immensely proud of it—it didn't mean he had to shoot every time and from everywhere. Donaldson often reminded him that when he began to shoot he lowered his head and could no longer see his teammates, who might be open and in an even better position to score. (Donaldson remembers watching the Calgary Flames on *Hockey Night in Canada* more than ten years later. On the screen, he saw Steve standing at the point; the puck came back to him, he wound up, stopped, and wristed a pass to a teammate in the corner. Donaldson jumped up out of his chair. "He listened," he shouted. It was a coach's moment.)

Donaldson coached Steve during the most pivotal minor hockey season for every kid—the year when body-checking is introduced (the age has changed over time, and varies depending on the province). There had been lots of body contact in the seasons before, of course, but it was incidental—two players going for the puck and bumping into each other—and (mostly) accidental. In Toronto, minor hockey tryouts are held in May for the next season, not long after the previous season has ended. When coaches select their team before that body-checking year, they have to ask themselves for the first time: *Is he big enough? Is he physical enough? Can he take it?* For those kids and their parents who get the good news, they spend the summer thinking about what's ahead, all of them excited, but many also fearful. Some kids mature earlier; some later. One of Donaldson's players made the team in May wearing size eight skates and started the season four months later in size elevens. Physical growth brings confidence and aggression to some kids, and gangly, mismatched body parts to others. And suddenly, on the ice, these kids are allowed to collide.

In that first checking year, players look like little bighorn rams running at each other, and they can't stop themselves from doing it. For weeks into the season and sometimes longer, the puck is almost forgotten. The kids have to prove—to their coaches, to their team-mates, to their opponents, and to themselves—that they are tough enough to hit and to take a hit, in this game, next game, every game. Kids who, up until this moment in their lives had been the best play-ers on the ice, who can skate like the wind, prefer suddenly to skate like the wind *without* the puck, to be the targeter not the target, the punisher not the punished. For some, the joy that has always been in their eyes is gone. And this is true of their parents, too; every season Donaldson saw it happen. A year or two later, and these kids are out of hockey, because of the financial cost of the game, because of the time commitment it requires—that's what the kids and their parents will say. But really, for many, it's because of body-checking.

The coaches try to prepare young players for body-checking: how to position themselves, how to take a hit. Although Steve was the youngest on the team, he was strong, big enough, solid on his skates, and had a "knack for hitting," as Donaldson puts it. Just before the moment of contact, when both kids were loading up for the blow, Steve would deliver his a split second sooner. He had this sense of timing. Maybe it was from having an older brother, and all the wrestling and banging around they did. Steve wasn't mean, Donaldson recalls, but he could really sting you.

For kids in minor hockey, there is "BC" and "AC"—"before contact" and "after contact." And there are the "BSs" and the "ASs"—the "before stars" and the "after stars." Steve, who had never been the best on the ice, at this moment began to see some of the best fall away, and he was still left standing.

One of Steve's teammates on the Braves was Mike Gardner. Both he and Steve were defencemen, and both played the same defence-first style of game. They both liked to hit, but Steve hit the hardest, "like a truck," Mike remembers. Steve was the best skater and had the hardest slapshot. Other guys with big shots blasted away from the point, the puck going high and everywhere, their teammates scattering from the front of the net to safety. Steve's shot, though he used it too frequently, was heavy and low, and nearly always on the net—just right for deflections and rebounds. But it was Steve's attitude that made him different. He was competitive, fearless, and intimidating to play against. You couldn't get to him, Mike remembers. If other guys got knocked down they would get embarrassed and shrink away; Steve would get back up as if nothing had happened. If he got into a fight and lost, it was as if he hadn't lost. He just kept going. If he tried to do too much and got caught out, which happened not infrequently, he'd try harder the next time. He'd do what needed to be done. The team might be behind 3–1, and in between periods "you'd look around the dressing room," Mike recalls, "and some guys, you know, they'd figure we've got no chance. I'd look at Steve and know we did.

"He was my idol," Mike says now.

Mike's parents and the Montadors were friends. The year before, Mike's older brother, Jimmy, had been killed in a bicycle accident; Mike was devastated. He and Steve didn't know each other well at the time. Maybe Paul and Donna said something to

Steve about the accident, Mike thinks, or maybe Steve just knew, but that summer after his brother's death, Steve hardly left his side. They had sleepovers, went for bike rides, played endless games of catch. "How many kids would do that?" Mike thinks now. When you are that age you don't even know what to *say* when someone dies, let alone what to do. It's not that they talked about Jimmy, Mike recalls. It was the comfort he felt from Steve being around, and doing things together.

When Mike thinks back on that year's Braves team, he remembers "a great bunch of guys. We lived in the same neighbourhood, went to the same school. Everybody knew each other. We were a real team. We felt like a team. We played like a team. About 80 per cent of us still see each other today." They play in a Braves reunion game every year; they play golf together. "It was the best team I ever played on."

"Steve just loved to play," Donaldson remembers. "He loved every part of it. He didn't just love it when he was a star, when body contact came in, when he had a coach who liked him and put him on the power play; when the team was winning. He loved it, every bit of it, all the time. You just knew *somehow* he would make it."

It's tough being a minor hockey coach. It's the long hours; it's all the rest of your life that you don't have time for. But it's the responsibility you have, more than anything. That's the best and the hardest part. These players like each other. They develop a feeling for each other and a loyalty to the team; they make a commitment to each other. In the end, they would do almost anything for each other. Those who are coaches and decision-makers in the league know this, but sometimes they forget. If they ask their players to go through a wall for their team, these kids might just try. They will do things that are good for themselves and for each other, and they will do things that aren't. Coaches and the other adults in the league have to know the difference between what seems to be good for these kids now and what might be bad for them later.

Donaldson himself had been a heart-and-soul player. He did some things as a player that, now as an adult, he knows that he shouldn't have done. As a coach, he came to know there were limits to what he should ask of his players—because each of these kids had his own life, and lots of years to live it.

At the end of the season with the Braves, Donaldson gathered the parents together. It had been a good year. The team had almost won the AA championship. The kids liked each other. The parents liked each other. Paul recalls how, at this point, Donaldson grew impassioned. There was no need to go on to AAA, he told them. They could stay together and get everything out of the minor hockey experience that they, as parents, had always said minor hockey was for. But the pull of moving up was greater, and several players left—including Steve, who would go on to play AAA hockey for the Marlies.

During that season with the Braves, at some moment that didn't seem to matter at the time—and in a way he later wouldn't remember—at age twelve, Steve got his first concussion.

CHAPTER FIVE

It took fifty years and the entry of Wayne Gretzky into the NHL for the possibilities of the forward pass to be realized. Passing had been at the centre of the Soviet game, but the Soviets were newcomers to hockey. Canadians were hockey's creators—and we were the best—so our way of playing, not theirs, must surely be the best. It only followed. The Soviets had looked impressive during the Summit Series and in their smattering of games against NHL players during the rest of the 1970s, but faced with a full NHL season, Soviet players would never be able to stand up to the test. Their way wouldn't work, Canadians knew, because it wouldn't work.

Born in 1961, Gretzky grew up in Brantford, Ontario, as a hockey prodigy. He spent endless hours on a backyard rink his father had built, learning the game's basic skills and much more. He took the game that he saw on TV and learned to think it, imagine it, and make it his own. Lots of kids dominate when they are young, especially when they are among the oldest in their age category, because they are bigger, more practised, and more mature than their teammates and opponents. But when age-related advantage begins to matter less, they get shuffled back into a pack they aren't familiar with and don't know how to handle.

Gretzky was a little taller than most kids, but he was skinny. He was a quick skater, but his stride was choppy and awkward enough to make coaches and scouts wonder. But Gretzky, even then, was able to stand back from the hype around his game to see himself. He realized that he wouldn't be able to dominate in the way his hero Gordie Howe had done—by power and force. Gretzky was a good puck handler, but with the puck on his stick his opponents could converge on him, slow him down, bump him, outmuscle him, and make him ordinary. He needed open ice, and he would only get that if he passed to teammates, then skated to open ice himself so they, in turn, could pass to him. "I skate to where the puck is going, not where it has been," Gretzky famously said. In other words, it is the players *without* the puck who determine where the puck is going, by going there themselves. This was how the Soviets played. For years, after the Summit Series, Canadian coaches and players had been able to ignore the Soviet philosophy, but Gretzky was one of our own.

Because of his own personal limitations, Gretzky uncovered other possibilities in himself, and he forced his teammates—Mark Messier, Jari Kurri, Glenn Anderson, Paul Coffey, even the Oilers' lesser skilled players—to do the same. By playing the way he did, he pushed them to skate faster, pass better, and shoot more accurately. And out of opportunity and necessity, they discovered they could.

In the 1980s, scoring in the NHL increased dramatically. Ordinary players compiled the point totals of superstars of the previous decades; outstanding goalies carried the stats of sieves. If passing were to be the focus, hockey's patterns had to change.

For its first century, hockey had moved in straight lines, forward and back. There was a *right* winger and a *left* winger, a *right* defenceman and a *left* defenceman. Those on the right stayed on the right; those on the left stayed on the left. Only the centre roved from side to side. For a winger, there was little worse than having a coach admonish, "Stay on your wing! Stay on your wing!" Yet with blue lines, a red line, and an offside rule which those generated, forward passes were difficult to complete. A pass in your own defensive zone was often too risky; in the offensive zone it was in territory too congested to succeed. Only the neutral zone, between the blue lines, offered promise—less risk, more opportunity—but there, players quickly ran out of space or needed to slow down, both of which defeated their purpose. To pass successfully, therefore, players needed to go to *wherever* there was open ice—on the right, on the left, it didn't matter—and skating diagonally across the ice offered a player more space to pick up speed without going offside. With Gretzky and the Russians, hockey changed from an end-to-end "north–south" game, to a side-to-side "east–west" one; from a game focused on the puck-carrier to one focused on everyone else.

A player wearing skates on a surface of ice had always had the capacity to go fast. The rule that prevented a forward pass had slowed him down; the rule change that allowed such a pass gave him the opportunity to go faster. The rule preventing substitutions, and the rule change that allowed them, had similar effects. A regulation that increased the number of players permitted to dress for a game enabled each player to play fewer minutes—and so enabled them to put more into each minute, to go faster, creating the need to be in better shape, to train off-ice and off-season in order to go faster still. Even Broad Street Bully–style intimidation as team strategy couldn't slow down this evolving game.

Those who grew up in the 1950s remember that decade as the NHL's tough old days, "when hockey was hockey and men were men." But they are mistaken. In the 1950s, the team that led the league in fights each season averaged only about one fight every three and a half games. The "Big, Bad Bruins" in the late 1960s pushed that number to slightly more than one every two games; the Flyers to one fight per game a few years later. By that time, the league's other teams—at first just trying to survive against the Flyers—learned to fight back. In the 1978–79 season, Philadelphia led the league with 74 fights, but Detroit had 71, Toronto 68, Chicago 66, Washington 64, St. Louis 62, and Boston 60. Of the league's seventeen teams, twelve had 50 fights or more. These numbers increased again during the high-scoring Gretzky years, peaking in 1986. That season, Detroit had 154 fights, almost two per game.

More offence usually leads quickly to more defence, but in the 1980s it led only to more offence. The players, realizing they could score, suddenly seemed addicted to scoring—and their coaches seemed powerless to stop them from trying. Even the goalies had no answer, until they began to increase the size of their equipment, but that came later. It wasn't until the 1990s that coaches and players started to adapt to the high-scoring game, and they did it with more hooking and holding, more clutching, grabbing, and obstruction—and for some, a team strategy called the "neutral zone trap"—to slow the game down. But the players were getting bigger; and the game, whether by passing or by forechecking, was always on the go. Collisions became more frequent and forceful. And while helmets, in an attempt to bring greater safety to the more dangerous game, had become mandatory for every new player in the league in 1979, they would turn out to have little effect.

Even with Gretzky's arrival in the NHL, the early 1980s were an insecure time for Canadian hockey. There was more fighting; the game seemed messy and out of control. In Gretzky's first season, only 4 per cent of the NHL's players were developed in Europe—twenty players were from Sweden, five from Finland, two from Czechoslovakia—but sixty-nine were now from the U.S., more than 10 per cent of the league. There were no Soviet players—they were still winning championships, believing in something bigger, and solidly behind the Wall—but their presence was felt everywhere. In February 1979, they had defeated the NHL All-Stars (a team that included three Swedes) in the three-game series known as the Challenge Cup. Two years later, they obliterated Team Canada, 8–1, in the final of the Canada Cup in Montreal. The Soviets, with their off-ice and off-season training, motivated by dedication not money, seemed to be going where the future was headed.

Canada seemed stuck—as a hockey power, and even as a country. After the heady years that had followed its centennial in 1967, and the early promise of Pierre Trudeau's government, Canada, it appeared, had returned to its historic backwater, and nothing embodied the country's disappointment in itself more than Canadian hockey. This was the perfect moment for Don Cherry.

Cherry had grown up in a working-class family, developed into a player of working-class skills, and become the coach of a team with a proud working-class identity, the Boston Bruins. As a coach, he was successful, cranky, and funny: a character; a blustering, ranting, syntax-challenged street kid who spoke basic truths. That was his schtick. After leaving the Bruins in 1979 for a big-money job to coach a bad team, the Colorado Rockies, he discovered that he couldn't live with losing, got fired, and moved from coaching to what became his Saturday night soapbox, "Coach's Corner" on

CBC's *Hockey Night in Canada*. It was 1982. In all the decades of radio and television before, it had been the play-by-play announcers—principally Foster Hewitt and Danny Gallivan—who had defined hockey broadcasts. Now it was Don Cherry. More than thirty-five years later, it is still Cherry.

Many Canadian hockey commentators at the time were dumping on Canadian hockey; many Canadian opinion-makers were dumping on Canada. But if you're a working-class kid, whining doesn't work and irony is too clever; pride is all you have going for you, and Cherry was proud. He spoke to the "deep in the bone" feelings of Canadians—the pride they had for their kids and families, their communities and hometowns; their loyalty to, solidarity with, and belief in their companies, organizations, and teams, in hard work and doing things together, in the jersey and the uniform and everything emblematic of the country. While others trashed Canadian hockey, Cherry stood up for it, and he struck a chord. His hockey was the stereotypical Canadian game of the time: all stitches and missing teeth, and rough and tough players; hockey that was "Rock'em Sock'em," as he called his bestselling annual videos; a game of the heart and spirit, not of science and the brain.

But Cherry also struck a reverse chord. Many people hated him. To them, his view of Canada was unsophisticated and simplistic, too much of a time that no longer existed. A Canada of Saturday-morning-at-the-hockey-rink/Tim-Hortons-coffee-drinking Canadians—not of the highly educated, eloquent, global moderns his critics were and wanted Canadians to be. Cherry embarrassed them.

In his early years on "Coach's Corner," European players and the way they played were more threat than reality. It wasn't until the early 1990s, after the fall of the Berlin Wall, that the Russians arrived in the NHL in important numbers. Quickly—along with other European players—they became stars, and began winning the NHL's biggest individual awards—most valuable player; top point-getter, top goal scorer, top goaltender, top defenceman, most outstanding rookie—in numbers almost equal to Canadians. The Europeans had come from nowhere to a very significant somewhere in the NHL, and to many Canadians this trend was very disturbing. Cherry fought back even harder than he had before. To him, everything about Canadian players was right; everything about European players was wrong. Good guys can only do good things; bad guys can only do bad. And loyalty means friends are always right.

Cherry overstated his own case, and he probably knew it. He had always respected talent—his favourite player was Bobby Orr, and as much as he had reason to hate the Montreal Canadiens, he didn't, because he knew that as talented as they were, like Orr, they were tough, too. And like other lunch-pail Canadian players he loves, Richard, Howe, and Gretzky are also Canadian. Cherry's fans accepted much of what he said about European players, laughed and rolled their eyes at the rest, and loved him all the same. They knew that as much as his rants seemed to be about Canadian hockey, they were really about Canada. Don Cherry believed in Canadian hockey, and he believed in Canada even more.

Don Cherry began on *Hockey Night in Canada* in 1980, the same year the Montadors bought their house on Avonbridge Drive, in Mississauga, the same city Cherry moved to two years later. Steve began playing hockey during Cherry's early years on "Coach's Corner"—in the basement family room, on the backyard creek, and in the circle at the end of his street. Steve grew up to embody the ethic and spirit that Cherry espoused. He was tough, hard-working, and hard-trying; he stood up for his teammates and himself; he did whatever his team needed whenever it needed to be done; he never backed down, never quit. He was under-noticed and underappreciated. He was an upper-middle-class kid and a working-class player. He was an underdog. He was the kind of guy Don Cherry loved.

Athletes are willing to commit *everything*. It's what being an athlete is. Hockey players had always been willing to commit more, and Canadian hockey players the most of all. They had a reputation and an identity to live up to, and they had their pride. But by the 1980s, NHL players were also being paid masses of money. They didn't need to bleed Leafs blue or Canadiens red—except for the years they were under contract to the Leafs or Canadiens. Then, as free agents, they could bleed Flyers orange or Penguins black. Fans had always given their all to their team; with all this money, they were less sure that the players were doing the same, and they were looking for evidence of some equal dedication. What defined *everything* continued to escalate.

In the 1980s and early 1990s, as Steve was growing up in Mississauga and playing minor hockey, *everything* for an NHL player was eighty games and four rounds of playoffs that continued until the end of May (in 1992, the playoffs extended into June for the first time). *Everything* was 110 games, preseason included. It was off-ice and off-season training. With more teams and more players, it was a longer career, more money, and more wear and tear. It was still injuries to the face—cuts and lost teeth—but more damagingly, because of the greater speed and increased collisions, it was injuries to shoulders and especially to knees. Once retired, this generation of players would be undergoing knee and hip replacements in much higher numbers, and at a much younger age.

Equipment was becoming more protective. After all, why should a player feel the pain of the ice, the boards, the glass, a puck, a stick, or an opponent's body if he didn't have to? Pain distracts and slows a player down. So why not play with abandon? With better equipment, why not hit with the same force with less pain, and hit more often; or with greater force but with the same pain? Why not do everything that is in you, with no compromise of performance for safety?

Head injuries, except for visible ones to the face, didn't seem much of a problem even at this time—and almost every player now wore a helmet. Some helmet-skeptics had argued for decades that head protection made a player more, not less, vulnerable. Without a helmet, a player can sense danger even from behind; with a helmet, it's as if his radar is jammed. But most of the skeptics just didn't like helmets. Fans identify with their heroes and want to feel for them, but when everyone wears a helmet, everyone looks alike, they argued. Fans can't see that look of triumph or disappointment on a player's face. When Guy Lafleur started up ice with the puck, fans could see his hair flip up then stream out behind. They could *feel* his speed. With a helmet on, some of that excitement was gone.

The helmet would have much less of a protective effect than everyone imagined, a fact that wouldn't come to be known until recent years. Helmets lessen the risk of a fractured skull, but do almost nothing to prevent concussions. Hockey still remained a compromise between performance and safety—even if most believed that both were possible pursued to their fullest, and compromise was no longer necessary.

Fighting had changed, too, but again our collective memory is dangerously wrong. The legendary tough guys of the 1950s— Gordie Howe, Lou Fontinato—fought a total of only four or five times *a season*—and that's against all opponents, not just against each other. The fighters of the 1960s—John Ferguson, "Terrible Ted" Green, and Reggie Fleming—fought seven or eight times a season, again *in total*. But in the 1970s, players like Dave "the Hammer" Schultz, Garry Howatt, and "Tiger" Williams routinely fought twenty-five times or more each year (Williams had thirty-four fights in the 1977–78 season).

One of Howe's unique legacies as a player is commemorated in a special category of achievement, the "Gordie Howe hat trick"—a goal, an assist, and a fight, all recorded in a single game. Including regular season and playoff games, and his years in both the NHL and WHA, Howe scored nearly 1,100 goals, had over 2,400 penalty minutes, and played more than 2,400 games. In that time, Gordie Howe had *two* Gordie Howe hat tricks.

Before the 1970s, the game was played to a different set of understandings. NHL players looked to earn a reputation for toughness *so they didn't have to fight*. That changed dramatically with the Broad Street Bullies. The league's total fighting numbers stayed at similar high levels in the 1980s, but more player-fighters contended for the league's lead in fights. Yet, with a few exceptions, they were still light-heavyweights—players who had to become good enough *as players* to get onto the ice to fight. But for fighters, giving *everything* was about to require a whole lot more.

Players had always needed to win, but in recent decades, they needed to win even more. The stakes were even higher. NHL franchises cost more money to buy; they were no longer just "hobby" investments for the local rich, but a significant percentage of their financial holdings. Most franchises were not located in traditional hockey areas. They had local competition from other sports teams and other activities. Spectacular success and spectacular failure were both possibilities. Serious money could be won or lost. The players became more important and more vulnerable. *Everything*, in terms of winning, had escalated as well.

In the 1980s and early 1990s, there was little to hold hockey back. The impact of the big changes that had been implemented throughout the century was just ahead.

In February 1993, Gary Bettman became the first commissioner of the NHL.

CHAPTER SIX

Every player, coach, and parent involved in peewee hockey everywhere in the world had their mind on *Le Tournoi International de Hockey Pee-wee de Québec* from the moment their teams were put together. Wayne Gretzky had played in the Quebec tournament. So did Mario Lemieux, Guy Lafleur, Marcel Dionne, Mike Bossy, Gilbert Perreault, Brett Hull, Steve Yzerman, Patrick Roy, and hundreds of other NHLers. All the best thirteen-year-olds would be there. Quebec was an opportunity for them and for their parents to see how they measured up against every other kid; to see how they were doing on their way to the dream. For these kids and their parents *together*, it was also a chance to get a taste of what that dream might be like: playing in Le Colisée, the arena that Jean Béliveau built, the home of the NHL's Quebec Nordiques, filled with NHL-size crowds making NHL-decibel-level noise, the national anthem played before every game; around the city, the kids in their team jackets, local people pointing, whispering, and teenage girls asking for their autographs. When Montreal Canadiens Hall of Famer

Steve Shutt was asked when he knew he wanted to be an NHL player, he said, "In Quebec," his eyes lighting up. It was when he walked down the dark hallway from his team's dressing room and turned into the light, towards the ice. When "all I could see was a wall of people. 'That's me!'"

Quebec was a hockey family's reward, and Paul, Donna, Chris, and Lindsay were all there. It was February, 1993. The Toronto Red Wings were the powerhouse team that year, led by Tom Kostopoulos, who would later play more than 600 NHL games. The Red Wings had beaten the Marlies several times during the Greater Toronto Hockey League (GTHL) regular season. Here, the two teams were in different brackets and wouldn't meet—until, perhaps, the playoff rounds.

The Marlies won their early games and played a team from Quebec in the group final. Le Colisée was packed. All the Marlies' parents, brothers, and sisters were cheering for the Marlies; all the thousands of others were cheering for the home team. Late in the second period, with the game tied, Steve had the puck in his own zone and the Quebec team was changing; there was open ice ahead of him. When he got to centre, the puck flipped over his stick. It took him a split second to notice. In that moment, a Quebec player grabbed the puck, skated in on a breakaway, and scored. Paul felt sick; Lindsay wanted to cry. Steve's head was down, his body slouched.

The game went on. Jim Nicoletti, the Marlies coach, put Steve back on the ice for his next regular shift. Immediately, Steve was bumping in the corners, moving the puck, jumping into the play, racing back, pushing himself to forget what had happened. Paul began to notice this only when the Marlies scored to tie the game once more; Steve was off the hook. Later, the Marlies scored again and won the game. It was a moment that none of the family ever forgot, and all of them remembered in their own way. For Paul, Steve had made an error, not a mistake—an error meant doing the right thing but not executing it properly; a mistake was doing the wrong thing for the wrong reasons. But most importantly for Paul, the team had picked Steve up. His teammates had helped him just as, at other times, he helped them. They had fought back together. It was the lesson of "team." Years later, Steve would play in the Ontario Hockey League (OHL) with the Erie Otters. In the team's program, as part of his player profile, he described the Quebec tournament as his most memorable moment in hockey.

The Marlies lost their next game, to the Red Wings, and were eliminated. The Red Wings then beat Syracuse in the final to win the overall championship. For ten days, Steve had been up against the best peewee hockey players in the world: Marián Hossa, Alex Tanguay, Brian Gionta, Simon Gagné, Mike Ribeiro, Tim Connolly, and Tom Kostopoulos among them. And on his own team, he had played with Daniel Tkaczuk. Steve had not been one of the tournament's best players, but it was now clear in his mind that he could play with them.

Steve returned to the Marlies the next year, but the following season he moved to another GTHL team, the Mississauga Senators. His Marlies coaches, after two years, saw him as a good, tough, hard-trying player. The coach of the Senators told Paul that he thought Steve had the talent to play in more critical situations as well—on power plays and penalty kills—and with the Senators he would be given that chance. But in his first season with Mississauga, nothing much changed. Other players were still more talented. Others still got more ice time in the high-skill moments of a game. But hockey's style of play was evolving. Teams were demanding players who were faster, tougher, and more competitive. They were demanding what Steve offered. If many players each year were still better, many more were worse, and Steve still loved to play. He remained a good teammate, and his team won most of its games.

Things were now a little easier for the Montador family. At home it was just Lindsay and Steve, and Lindsay—older and more independent—went to the rinks only when she wanted to. Chris was in Wilcox, Saskatchewan, attending Notre Dame College, which at the time was the best hockey school in Canada. Wendel Clark, Curtis Joseph, and Rod Brind'Amour had gone there; so would Vincent Lecavalier, Brad Richards, and many others who went on to play in the NHL. The rule in the Montador family had always been that school came first. But little by little, under the compromises of AAA hockey, the definition of "first" had begun to slide. It had become not much more than "if Chris/Steve is passing, even if it's by the skin of his teeth, he can play"—and at times Chris had been pretty close to the line. But at Notre Dame, he would have to earn his playing time for the Hounds in the classroom as well as on the ice.

After his season with the Senators, Steve got a chance to play Tier II Junior A with the St. Mike's Buzzers. It was a big step from bantam to junior. Steve had always played with and against kids his own age; in junior, the players might be four years older—bigger and stronger physically, more developed emotionally, intellectually, and as players. But St. Mike's, a hockey school even more legendary than Notre Dame, was too good a chance to pass up.

St. Mike's had been a central building block for the great Toronto Maple Leaf teams of the 1960s. This was before the NHL's universal draft was introduced with expansion in 1967. After the Second World War, NHL teams had begun building their farm systems. Players could be signed by NHL organizations at a young age to play for local teams that the big clubs sponsored. Many of the most promising of them were from small towns, especially in the north. As these young players got older, they needed to play for better teams in more competitive leagues, which were usually located in the south. But if these kids were to leave home and move to the big city, their parents wanted to be sure they were safe. That's where Catholic schools came in. What could be better for the welfare of their kids, parents believed, than to be looked after by priests? Detroit sent their young out-of-town players to Windsor, to play junior for the Spitfires and to attend Assumption College, a Catholic high school. The Leafs sent their prospects to St. Michael's College School (St. Mike's)—Tim Horton, Frank Mahovlich, Dick Duff, Dave Keon, Gerry Cheevers, and many more. Red Kelly and Ted Lindsay, who both later signed with Detroit, also attended St. Mike's. Next door to the school was the arena, and across the front of the arena, in big letters, a sign read, "Teach me goodness, discipline and knowledge."

St. Mike's had withdrawn from major junior hockey years earlier; during Steve's time there, in the mid-1990s, it had become a prime feeder to U.S. colleges offering hockey scholarships. Most of the team's players attended St. Mike's, and were still of high school age, eighteen and under, making St. Mike's the youngest team in the league. Steve—at fifteen, moving up from bantam and with his late birthdate—was its second-youngest player.

Kevin O'Flaherty and Steve were rookies together on the team. Kevin's grandfather, "Peanuts" O'Flaherty, had played with St. Mike's in the 1930s before having a pro career as a player, coach, and scout. Kevin's father, John, had played Junior B with Steve's minor peewee coach, Jim Donaldson. To Kevin, though Steve was a year younger than he was, Steve seemed the older one. It was the way he held himself, Kevin recalls. Steve seemed so comfortable in his own skin, right from the first day, yet without cockiness or swagger. The younger guys liked him because he wasn't intimidated by the older guys; the older guys liked him because he seemed like them. "Their mothers and sisters liked him too," Kevin remembers with a laugh. Yet Steve was no hotshot prospect, no "golden boy," and as a fifteen-year-old he had to prove to his teammates, coaches, and opponents—and to himself—that he could play at this level and handle everything it demanded of him. Against big, tough teams, sometimes in small-town arenas, against big, tough local heroes four years older than he was, he had to show he was tough enough.

Steve wasn't a fighter, Kevin remembers, but he fought because the team needed him to fight, so that others didn't have to. St. Mike's had another young player on the team, Kip Brennan, a big left winger. Brennan would go on to have a thirteen-year pro career; in his sixty-one NHL games with the Kings, Ducks, Islanders, and Atlanta Thrashers, he had one goal, one assist, and 222 penalty minutes (PIM). In forty games with St. Mike's, Brennan had 155 PIM; Steve had 145. Brennan was a heavyweight; Steve a light-heavyweight. Yet Kevin recalls, "I never saw fear in his eyes. Ever." Nor did Steve pick his spots and prey on the weakest, most vulnerable opponents. "Absolutely not," Kevin says. Taking on the weak would be about him; taking on the strong was about the team. It might seem that the toughest guys are those who deliver the biggest hits, with a shoulder or a fist. But the toughest guys are *really* those who are willing to *take* the biggest hits, to make a play to help their team. And Steve, Kevin explains, was truly tough.

They were both from Mississauga. Kevin attended a high school in the area; Steve went to St. Mike's itself, in Midtown, Toronto. Paul often dropped Steve at the Kipling subway station on his way to work, where Steve would catch a train to school. When Paul was away, Steve took a commuter train from Mississauga to Toronto's downtown Union Station, then transferred to the subway, sometimes carrying his sticks and hockey bag. For Steve, hockey was a big commitment: up at 6:00 a.m.; home at 7:00 p.m., or later when there was a game. Paul was now the head of Johnson & Johnson Canada, busier than ever and travelling more often. He made it to most of Steve's games—and when he couldn't, and when Donna was busy with Lindsay, Steve got a ride with the O'Flahertys.

Steve and Kevin didn't say much during these rides. On the way to the rink they had the game on their minds; on the way home, they were too zonked. Yet whether in the quiet of the car, or in the dressing room, or on the ice, Steve had a focus about him that Kevin remembers well. "He had this internal fire. You could see it in his eyes." There were lots of players better than Steve, "but he was so determined. He could overcome things. What was going to stop him?"

Steve had had two pivotal hockey moments in his life. First, the decision to play AAA; second, his experience in Quebec. And now, he had a third: the 1996 OHL draft. The best sixteen-year-old players in Ontario, in U.S. states east of the Mississippi River, and in Europe were eligible. This was another chance for kids and their parents to see how they stacked up.

Steve was selected in the third round by the North Bay Centennials, the 36th overall pick of the draft. Rico Fata, later Steve's teammate in Saint John, was the draft's first overall pick. Kip Brennan went in the first round, 4th overall; Bryan Allen, who played more than 700 NHL games, went 9th; Mark Bell was 15th; Manny Malhotra, 17th. Tom Kostopoulos, Steve's Red Wings rival, went in the second round. Others selected who would have NHL careers included Sean Avery, Paul Mara, Jason Williams, Mike Van Ryn, and Chris Neil. Steve's former Marlies teammate Dan Nicoletti, Jim's son, was also drafted, as were Mike Gardner, Steve's Mississauga Braves teammate and buddy, and Nick Robinson, who Steve would later get to know in Peterborough and who became a lifelong friend. In all, 291 players were taken—the best of the best. Only a handful of them made it to the NHL.

For Steve and Paul, this was decision time. Major junior players in Canada receive a small weekly stipend to play for their teams, many of which are highly profitable, but the stipend is large enough that the NCAA considers the players professionals and ineligible for scholarships and college competition. Steve had one more year of high school remaining. He could stay at St. Mike's, have another season of good competition, then graduate and take a U.S. scholarship; or he could leave home and go to North Bay, live with a billet family, play in a league with multi-hour bus rides, attend school under far from ideal circumstances, gain scholarship credits to help pay his future expenses to a Canadian university, and play against the best young players in the world.

To Paul, this was a very difficult decision. To Steve, it was not. School had always come first in his family because no parent or kid in any family had the right to think otherwise. But school had never been first for Steve. What drove him was elsewhere. Most of what he had learned, most of the people-lessons and life-lessons, had come from hockey. He had been in more tough situations, and had more responsibilities placed on his shoulders by the age of fifteen, than most people have in their lifetime. He'd had teammates who counted on him absolutely. He'd had an arena full of fans telling him he sucked. He had held up and met every challenge.

Paul talked to Steve. In his manager's way, he tried to encourage, not push, but Steve knew which way Paul was "encouraging." Still, Paul wanted to help Steve understand all of the implications of a decision that, in all probability, would affect the rest of his life. Paul wanted him to consider it carefully so that he could make the right choice himself. He did not want to intrude—or seem to intrude—on Steve's decision; if he did interfere, that might have its own lifetime of implications. Paul wanted Steve to understand that they were really on the same side, that this decision wasn't just about junior hockey or college hockey—it was about Steve's future.

Steve didn't say much, and Paul knew which way the wind was blowing. He also didn't have much stomach for the fight. This decision wasn't much different from all those times he had been offered a promotion, a new chance to learn and to move one more step up the ladder. And had he ever turned those offers down?

To make himself feel better, Paul made Steve promise that if he did go to North Bay, it would be on the condition that he stayed in school and graduated; and he made a promise to himself that if Steve did go, every time they spoke on the phone he would always ask him about school first, before hockey.

The two of them sat at the kitchen table. Paul knew that if he said no to Steve playing junior hockey, he'd have, as he puts it, "a very unhappy teenager on his hands." He wasn't prepared to have that happen. The horse had left the barn years earlier: hockey was what Steve did. And from that moment on, hockey was where he'd go, as long and as far as it took him.

Steve made the decision, not Paul. He went to North Bay.

CHAPTER SEVEN

It was early September 1996 when Chris got a call. It was Steve, in North Bay. "I should have taken the scholarship," he said immediately. He didn't even say "hi" first.

North Bay is a small city of about 50,000 on the shore of Lake Nipissing, near the southern edge of the Canadian Shield, an immense area of mineral resources, lakes, trees, winter cold, summer sunshine, wide open spaces, and not many people. The city calls itself the "Gateway to the North," but more than that it has been a transportation junction between east and west. By car, on the Trans-Canada Highway; by train, on the railroad line from Halifax to Vancouver; and, for centuries by canoe, on what is sometimes called "the first Trans-Canada Highway," a waterway that runs from Montreal, up the Ottawa River and the Mattawa River to Trout Lake, then over a seven-mile portage to Lake Nipissing, and down the French River to Georgian Bay and Lake Huron. North Bay's first senior hockey team was called the "Trappers"; and its first major junior team was named the "Centennials," coming to the city as it did in 1982, one hundred years after the railroad had arrived in North Bay.

The city had once been a centre for lumbering and mining, for sport-fishermen and hunters. RCAF Station North Bay was constructed in 1951, in the midst of the Cold War; a decade later, the North American Aerospace Defense Command (NORAD) built an operations centre sixty storeys deep into the rock of the Shield to monitor Canada's airspace and warn of a nuclear attack. The base became North Bay's largest employer. When the Cold War ended, its operations were cut back. By the time Steve arrived in the mid-1990s, the city was doing better economically than most places its size—especially in the north—but many of its young people Steve's age were leaving for the action and the jobs in larger cities in the south, like Toronto.

North Bay hockey had also seen better times. Most of hockey's promising young players were now being developed in the competitive hothouses of the suburbs to the south. The north's hockey advantages were long gone: cold weather, ice more months of the year, kids with fewer things to do and more time alone to develop a star's special skills. But no one played outdoors now, *especially* in the north—it was too cold—and the indoor game could be played twelve months a year. Hockey had also become less an individual game of great personal skill than a team game based on coaches and systems. Kids needed time to play with and against others who were just as good; they needed highly competitive time, not time alone. In the south, those players and teams were in the same cities. In the north, they were a long, time-wasting car ride away.

North Bay had had senior amateur hockey for decades, a minor professional team for a year, then senior again. It had built a new arena in 1955, seating 3,500; the rink, like the city, was too big for senior amateur hockey, too small for anything more—and the OHL was tethered by bus travel to southern and eastern Ontario. Then, in 1972, the league expanded to Sault Ste. Marie and Sudbury. Both cities were bigger than North Bay, but if distance was no longer the unanswerable question, then North Bay could dream.

In 1982, the OHL's Niagara Falls Flyers were looking to the city of Niagara Falls to build a new arena, and when the city refused, the Flyers moved to North Bay. Importantly, their coach moved with them. Bert Templeton had been successful coaching junior teams in Hamilton, St. Catharines, and, of course, Niagara Falls. He would continue that success in North Bay, with the Centennials never missing the playoffs in his twelve seasons, the team making it to the OHL finals twice, winning once and going to the Memorial Cup. But Templeton was looking for a raise after the team won the OHL title in 1994, and when he didn't get it, he left.

In the 1990s, Junior hockey was changing. NHL players were still mostly ineligible for the Olympics, and the World Championships, held at the same time as the Stanley Cup playoffs, were second-rate. For Canadians, the only defining international hockey competition was the World Junior Championships. Played in a post-Christmas, New Year lull in the NHL schedule, and broadcast on TV, these games generated a huge new appetite for junior hockey. And this enthusiasm came at a time when old local rinks were needing to be replaced. New ones went up—mini-NHL arenas with 5,000 comfortable seats, better food, and videoboards. The OHL continued to expand. Franchises increased in value.

With more money invested, owners began running teams like businesses. But North Bay had a new problem. It was the second-smallest city in the league. Much of its economy was government-based, in industries—military, education, health—that had little

77

sponsorship or promotional money to spend on a hockey team. The Cents played in an old, small rink and had little prospect of a new one.

Templeton had made North Bay's limited prospects work because he had made the Cents competitive. But the season after he left, the team lost in the first round of the playoffs. A year later, they missed them entirely. This is the team that Steve joined before the start of the 1996–97 season.

Steve's call to Chris came soon after he had moved to the city—before the cold had come, before the team's first five-game losing streak and seven-hour bus ride to Erie, before it was clear that this was a bad team going nowhere. As Chris puts it, this was an "I'm in freaking North Bay, four hours north of Toronto, and now I have to make the NHL because what the heck else am I going to do" call.

Then things began to pick up. Despite everything else, this was a major junior team, one of only forty-nine in the country. Except for some kids from Europe and a few top draft picks who had gone directly to the NHL, Steve was playing against the best under-twenty-year-olds in the world. He was going on road trips. In North Bay, he was playing in front of 2,000 people, near to the lowest attendance in the league, but that was 1,500 more than had ever watched him before, except in Quebec. And on the road, 4,000 or more might be at the games. The Cents were losing, but he was playing—and he was only sixteen years old. He was a big deal, and he was on his way to becoming a bigger one.

He stayed with a local family like all the out-of-town players did. Many of these families had billeted players for years. Most of them were hockey fans, and some enjoyed being almost second parents to teenagers who were tasting freedom for the first time but who needed a little guidance. For some players and their billet families, this would begin a lifelong relationship. But Steve's accommodations were "terrible," as Paul recalls. "There was peanut butter, and some frozen food in the freezer, and the people were never around." Steve said nothing about this to Paul and Donna; as the new kid in town he wasn't sure he had the right to complain, and besides, having mac and cheese for breakfast, lunch, and dinner didn't seem to Steve such a bad deal. When Paul and Donna came to visit him and saw the conditions he was living in, he was moved to a new billet, then to another, before finally coming to live with a family he liked—and one who gave him the freedom he was looking for. They let him bring teammates over to their house, and if that led to drinking at times, well that's what teammates and kids in small towns did. They brought him ice fishing on Lake Nipissing, and if that meant sipping a little whiskey to stay warm, well that was what ice fishermen did, too.

Steve had to get used to a new school. Chris had needed to do the same a few years earlier, but that was at Notre Dame, where if you didn't do your schoolwork, you didn't play hockey. In North Bay, the team didn't pay much attention to the players' schooling. Those who really wanted to go to school went; the others went sometimes. And while the school was proud to have the Cents as students, some of the teachers were not. The players were absent more often than other students, and at times had to leave class early; they got special treatment. When they were there, the other kids hovered around them; they disrupted the class.

It wasn't easy for Paul and Donna to know how Steve was doing. He was 400 kilometres away. This was before cellphones, email, texting, or FaceTime—and what could they really tell from a phone call a couple of times a week? Paul wasn't a talker. He would ask Steve about school, as he promised himself he would, but Steve was also a teenager, and he didn't say much either. During Paul and

Donna's occasional visits, sometimes with Chris and Lindsay in tow, it always seemed like Steve was either getting ready for a game, or playing a game, or winding down from a game, and all they could really tell about how he was doing was how he was doing on the ice. If the team had won or lost; if he was playing power-play minutes; if he looked good or right to their eyes.

His friends were his teammates. His classmates had Friday nights and weekends to be with each other; Steve had games to play, in North Bay or on the road. The team travelled more than 16,000 kilometres a season, the second-highest total in the OHL— all of it by bus, almost 200 hours in all. Hours playing cards, trash-talking, watching bus-movie classics like *Caddyshack* or *Dumb and Dumber* on the VCR, or napping before or sleeping fitfully after the game. These were Steve's scheduled hours; during them, he had no choices. The hardest hours were his unscheduled ones: after school, after practice, before the next game.

There was a bar they went to. Well, not really a bar, it was more of a barber shop, Chris recalls, or maybe a hair salon. It was a place the players could go, close the blinds, spend time with each other and with girls they'd invited, and drink. Chris had wondered what Steve did in all those hours he wasn't playing hockey. What was there to do in North Bay? Ice fishing and snowmobiling are fun, but how many times can you do that? North Bay might have been a great place to live for older people during their settled years, but what about for teenagers, especially if they were from the south, as most of the Cents players were.

It was the drinking that most surprised Chris. At home, he and Steve had snuck the occasional beer, as most kids do. But his years in Saskatchewan had been a real eye-opener for Chris. Playing weekend games in small towns, he had never seen so many people so wasted. He thought he had gotten used to it, but during one visit to North Bay he saw his brother and his buddies in action. "This isn't a place he should be," thought Chris.

Paul went to road games that were close to Toronto—and sometimes those farther away when he could find a reason to schedule business nearby. He recalls a game in Oshawa late in Steve's first season with the Cents. Steve was skating up the ice, without the puck, and out of nowhere an Oshawa player, Nathan Perrott, raced at him and crushed him against the boards. Steve didn't get up. Perrott was over 200 pounds and twenty years old. He would play 89 NHL games, mostly with the Leafs, recording four goals and five assists, and 251 PIM. Perrott got a penalty. Two or three rows in front of Paul, before Steve even got up from the ice, an elderly woman in a hand-knitted Bobby Orr sweater yelled, "Hit him again, Nathan."

Steve was helped to the dressing room. A short time later, he was back on the bench, then on the ice. He skated over to Perrott and "tapped him," signalling that he wanted to fight. Perrott was a fighter and, as Paul said, "Steve was a wannabe."

"He acquitted himself fairly well," Paul recalls, "and may have impressed his teammates. Nobody told him to do this. He was going to stand up for himself. I remember thinking, 'He's not a kid playing hockey at home anymore. He's part of a different world now.'"

Midway through the year, the Centennials coach, Shane Parker, was fired and replaced by Greg Bignell, but not much changed. The team finished the season with an embarrassing 14–44–8 record, and were out of the playoffs for the second straight year. Steve had gotten a year older, and a year better in some ways. But he played on a dysfunctional team in a not-always-constructive environment. When school was done, Steve went home.

Home for him was now in a different place. The Montadors had moved to Lakefield, a picture-perfect town of old stone houses and churches on the Otonabee River, fifteen kilometres north of Peterborough. Paul was the head of Johnson & Johnson Canada, a division of one of the world's largest health care companies that had begun its operations in Peterborough three decades earlier, expanded, then contracted as manufacturing was moved to less costly sites offshore. Peterborough had been able to attract large companies in its past—Canadian General Electric, Quaker Oats, Outboard Marine—initially because of the availability of local sources of hydroelectric power, and later because of the lifestyle that the area afforded. Executives of a company who were offered a more senior job might not accept the promotion if it were to a small town. But if it were to Peterborough, they might. With its lakes, rivers, and proximity to Toronto, Peterborough was a blue-collar town that appealed to white-collar people.

Johnson & Johnson in the U.S. had been a major supporter of an injury-prevention organization called ThinkFirst, whose purpose was to get people, especially kids, to "think first" before they dove from a cliff or played chicken with a passing train. Spinal cord injuries were the group's initial focus; head injuries and concussions came later. Because the U.S. head office had made ThinkFirst a priority, Paul decided to do the same in Canada. He would be on its board for more than fifteen years. The head of ThinkFirst Canada was Dr. Charles Tator, a neurosurgeon. Twenty years later, it was Dr. Lili-Naz Hazrati from Tator's sports-concussion group who examined Steve's brain and discovered his CTE.

Peterborough is a hockey town. The Petes had arrived in the city in 1956 and, with a history of great players—Steve Yzerman, Bob Gainey, Chris Pronger—and even greater coaches—Scotty

Bowman, Roger Neilson, Mike Keenan—they had been at or near the top of the OHL since then. If you were a boy growing up in Peterborough, you likely played hockey; and if you played hockey, you likely also played lacrosse, as Peterborough's top lacrosse team—at different times called the Lakers, Trailermen, Petes, and Red Oaks—had won many Canadian championships. And if you were born in Peterborough, you were likely raised in Peterborough, and got married and had kids in Peterborough. Summer cottagers came and went by the thousands; Peterborough's 80,000 year-round residents stayed. For those new to the city, it wasn't easy to fit in.

When Steve was twelve, some Peterborough and Toronto fathers put together a team that included kids from both cities to play in summer hockey tournaments. Steve was one of the Toronto players on the team, Jay Legault one of Peterborough's. It is where they first met. The team, called the Trent Stars, offered good competition over the summer and took up only three of their weekends. There were no practices. The experience was especially useful for the Peterborough kids. Toronto was the big city—its best teams recruited players from across the entire GTA, which had fifty times the population of Peterborough. During the regular winter season, when Peterborough teams played Toronto teams at tournaments, the Peterborough kids thought they would get killed. The Trent Stars were a chance for them to change in the same dressing room as the Toronto kids, go out onto the same ice surface, and see for themselves. They would have to raise the level of their game. The Stars played together for three summers—and the Peterborough kids discovered, of course, that they weren't so bad after all.

The first night Steve was in Lakefield, he looked through the Peterborough phone book and found the name he wanted. "Hey,

Jay, it's Steve Montador," he said, "and we've moved into the area."
It was a Friday or Saturday night, Jay recalls: "Right away, I got my
dad to drive me out there to pick up Monty. I introduced him to all
my hockey friends and to the other Peterborough guys who played
in the OHL."

This was the summer, and the Peterborough "hockey guys"
were back together again, picking up where they had left off the
September before—but now with a season's worth of stories they
couldn't wait to tell. Nick Robinson had been in the Soo, Chad
Cavanagh in Sudbury, Ryan Ready in Belleville, Jay in Oshawa,
and Colin Beardsmore with Steve in North Bay. Also in the group
were some players who weren't as good or as interested in hockey,
who had played locally on Junior B or Junior C teams. One of them
was Mike Keating, or "Keats" as everyone called him.

Jay, Steve, Keats, Nick, and some others worked as on-ice
instructors at a hockey school. It ran for seven weeks. If they were
lucky, they got four shifts a week, and made between three and four
hundred dollars. They also picked up odd jobs to fill in their other
days. Many of their parents had small businesses in the area. They
cut wood for Nick's dad, who was an arborist. They roofed houses,
poured concrete, and put up tents for Lester Awnings. When they
weren't working, they hung out and trained.

The OHL players, all with at least one junior season behind
them, were now more serious. An NHL future had always been the
dream. Now they knew what they were up against. Now they could
feel time running out. The next few years would determine whether
they would make it or not. They got together at the Peterborough
Lift Lock and ran; they went to the gym; they pushed each other.
They rented the ice to work on their new moves and to laugh at last
year's ones that still didn't work. On the ice with them were the

Peterborough pro-hockey guys—Cory Stillman, Marc Savard, Dave Reid, and others—the guys who had made it. Some of them had been born in the city; others had played with the Petes, had loved the atmosphere of its winning team and adoring fans and, with pro money in their pockets, had come back: to jobs in the city, to summer places on one of the lakes, to marry a girl they had met in junior, to make Peterborough home. They all knew each other— the pros and the juniors. They were the same guys, only a few years apart, and on the ice the pros had to live up to their local reputations, and the juniors had to show them they belonged. For all of them, it was a chance to work up a sweat and a thirst.

For Steve, Jay, Keats, Nick, and the others, a job instructing at the hockey school, roofing houses, or pouring concrete was all about having money for the weekend. None of them had places of their own; none of them wanted to invite twenty-five occasionally out-of-control guys over to their parents' house. So they had "field parties," as they called them. At the Lagoon or at Jackson Park, some place enough off the beaten path that they could make a little noise, drink underage, not bother anybody, and not have anybody bother them.

They also took things a step further, and built a cabin in the woods between Douro and Norwood, and called it "the Shack." The father of one of their friends was a dairy farmer. He had a little land he wasn't using, supplied them with some barn lumber, and helped them build it. All of them pitched in. The cabin was about fifteen feet by twenty feet, with one room on the main floor and a ladder that connected it to two mini-lofts that could sleep four or five comfortably, and which often slept eight. On weekend summer nights, there might be thirty or forty people around. Everyone brought a case of beer, and they would have a bonfire, hang out, talk, and act foolish, as Jay puts it. They would "chirp" and needle and get under each other's skin, and everything was fair game. The only offside words were the ones that weren't funny.

They talked a lot about girls, and at two in the morning, they talked about things that sounded profound. But it didn't matter what they talked about. This was about being with each other. They were seventeen, eighteen, nineteen years old, and at the Shack they could do what they wanted to do. "We spent one whole summer building that thing," Jay recalls. "Keats, Monty, Nick, all the guys. It was our baby."

Steve fit right in with the Peterborough guys because of hockey, drinking beer, and girls; because of his humour and the way his mind worked; because he was willing to try anything and do anything, look stupid and be stupid if that's what the moment was about. He wasn't shy about the "shenanigans" they got up to, as Keats calls them, they were just more fodder for stories and laughs they would share years later. Steve fit in because he never thought of himself as a hotshot. He lived in the big house on the lake. He was a Toronto guy, but he was different from every other Toronto guy his Peterborough friends had met. Toronto guys acted as if they knew something Peterborough guys didn't. Steve wasn't like that. He was one of them. He liked to hang around, pitch in, and do what they did, as if what they were doing mattered, as if Peterborough mattered, as if they mattered. Even when he made it to the NHL.

"He gave everyone the time of day," Keats recalls. "He used to say, 'You never know what a person has to say, so why not listen?'"

In time, for the Peterborough guys, their year-round lives took over their summer lives. They got married, had kids, got adult jobs. In time, the drinking got to be too much; the stupid stuff too stupid. Keats, Jay, and Nick went away for a few years. Keats and Jay came back; Nick stayed away and got a job in New York with a large private investment firm, but he came back to visit his family. When Paul and Donna later returned to Toronto, only Steve could pack up and walk away. Yet he always came back, too, for a few days each summer—and Keats, Jay, and Nick visited him wherever his NHL career or summer freedom took him. Between these times, Steve kept in touch with them by phone; later, by email and text. When they were on his mind, he let them know. Just a few funny, familiar lines. It made them feel like they were always close. Jay, Keats, and Nick began as Steve's summer team, and became his lifelong team.

CHAPTER EIGHT

The next season in North Bay was much like the first. Still, it had its moments. The Cents had a game early in the year against the Sudbury Wolves. Steve Valiquette, six-foot-six and twenty years old, was the Wolves' regular goalie, but this being a game against North Bay, he was given the night off. So "Vally," as everyone called him, decided to milk the moment. He sat back contentedly at the end of the Wolves' bench, a big smile on his face, a white towel wrapped around his neck. He and Steve had never met. Steve lined up for a faceoff near the Wolves' bench and looked over at Vally. "What's the matter, Fatso? You not playing tonight?" he said. Vally, everyone knew, was not the fittest athlete on the planet. "Let me have that towel to wipe my brow." The puck dropped, Steve jumped into the action, and Vally was left stunned but smiling.

Later in the season, Vally was traded to the Erie Otters, who were in a desperate chase for the playoffs. Not long afterwards, the Cents' management, realizing that the team was going nowhere, began dismantling it to build for a new future. Steve was also sent

89

to Erie. The day after Steve was traded, the Otters were having a team breakfast before getting on their bus to head out onto the road. Some of the players decided to go for a walk, Vally included. Suddenly Vally looked up, and Steve was walking beside him. Vally had skipped breakfast and hadn't heard the news. He couldn't believe what he saw. "Hey, Gillies," Vally blurted. "What're you doing here?" Trevor Gillies had been the second-round pick of North Bay in Steve's draft year. Vally looked again. "Hey, you're not Gillies. You're the guy who wanted my towel." Steve would become, as Vally put it, "my best friend in hockey."

For Steve, the move to Erie meant a new city, new country, new school, new classmates, new courses; and also a new team, new teammates, and new possibilities. But the same old bus rides. He had gone from almost the northernmost team in the OHL to the southernmost, and the Otters' nearest opponent was two hours away, with Lake Erie's "lake effect" snow in the way. When their team walk was done, Vally and Steve got onto the bus and each went to where each was supposed to go. Vally, an overager with seniority headed to the prize spot on the bus—"the Pit," where the second-last row of seats had been turned around to face the back row, and where the card games were played. Steve, not much more than a rookie to his new teammates, stood at the front, waited for the other players to take their seats, found an empty one, and sat down.

Life on the bus is about real estate, Vally describes. And it's about rank. The older players have it; the younger ones don't. The older players sit at the back, eat first, and choose the movies. The younger players sit at the front and wait. But what they all did together was chew tobacco. Chewing tobacco was the bus's pastime. "We couldn't smoke cigarettes," Vally explains. "We were hockey players. We had spittoons all over the place. We bought Skoal, everybody did, and Cherry was the flavor *du jour*, until it got old, then Wintergreen, which was lousy, then Mint Straight." The players were each paid $40 a week by the team; the team held back half—or $19.79, to be exact. "Our paycheques were for $20.21," Vally recalls. For Steve, "that was gone in two days. I always thought he was from a family of no means." With post-game drinks, a late-night run to Subway, and tins of chew at four or five bucks a piece—one tin every day and a half—the money went fast, and Steve's went the fastest. A player chewed even when he didn't want to chew, because, as Vally puts it, "you wanted to be an OHL player." With dip in their mouth, playing cards with their buddies, on that bus they were kings of the world.

Steve would always run out. "Can I borrow a chew?" he'd ask Vally. "It was always 'borrow,'" Vally says, "like he was going to give it back. Then, on the very rare occasion when I'd ask for some of his, he'd say, 'Sure, but it's recycled.' He'd have taken a chew that he wasn't completely finished with, and put it back in the tin, to save it for later."

After the trades, the Otters went on a roll. Friday night, a win; Saturday night, a win; Sunday, a day off; Monday, it all started back up again. And the more they won, the harder they worked; the more they won, the more fun they had; and still the more they won. Week after week. Everybody's a great guy when you win.

Earlier in the season, the Otters were just missing a few pieces, but often trades don't work. Trading wisdom says that while you should always know what you're giving up, you can never know what you're getting. You know on paper, you know what your scouts tell you, but will that new player fit in? He was a leader on his old team, perhaps. Part of what made him so good was the responsibility he had, the respect he held, the specialness he felt. Is that same role open to him here? And what about the player who was filling that role before? Can he adapt? And all the time, there's a game tomorrow, and the playoffs are looming. But in Erie, Steve and Vally fit in.

Dale Dunbar was their coach. He practised the team hard, and had them do push-up and sit-up circuits in the gym and do their time in the weight room. Everything was structured. They had lots of team meetings. The Otters bought them each their own director's chair to sit in, their own "Road to Success" T-shirt, their own "beautiful robe," as Vally describes it, with their own number on it. They even had team flip-flops. There was energy and purpose to everything they did, and Steve loved that. He loved the feeling of being a part of something. For the younger players, Steve was their "sheep herder," says Vally. "They needed someone popular with them who was also cool with the vets. He was the captain of the rookies. He got everybody going in the same direction." The Otters began to feel like a team.

Other players used practice to prepare, Vally recalls. But Steve prepared for practice. "He had this attention to detail." He was such a competitor; games, practices, it didn't matter, and as a goalie, Vally loved playing behind him. "He didn't skate like anyone else," Vally says. "He skated *into* the ice. You could hear it. He went deeper. He had power. There was commitment." While for many, defence is a grim position to play, with opponents always running at you from behind, "Monty absolutely loved playing defence. He went in first after the puck every time. And I knew, any fifty-fifty puck was his."

The Otters' roll continued. Another Friday win, another Saturday win; and a Saturday night after a win was always the best. The week's work was done. The team would go to a bar called the

Magic Lantern on the outskirts of town. Pennsylvania's drinking age was twenty-one, but the players knew that they would be served at the Magic Lantern. There were no women there; Erie's women could wait for them a few hours longer. "We would just sit there, drink warm pitchers of beer, and be the happiest guys on earth," Vally remembers. "They had a jukebox, we'd play music and tell stories, and Monty's were always off the charts; about some girl he'd met. Then you'd get up and be having a piss, and you could still hear that laugh of his all across the bar."

By Saturday night, of course, Steve would be four-days broke. "I loved him so much, I paid his way through junior," says Vally with a laugh. He has a story about visiting Steve the following summer in Lakefield: "I'm driving down his street, and I know it's his street because I've got his address. But all I see are these mansions right on the lake. Mansions! Then the street ends; it's a dead end. So I turn around and go back to town. This is before cellphones, so I call him from a pay phone. 'Hey Vally, what are you doing?' he says. 'I saw you drive right by.' Well, I get to his house and I lay into him. 'Monty, you've got to be kidding me. You're rich. You never had five bucks in your pocket and your dad's got a Porsche Boxster.' It was the most beautiful house I'd ever seen, with this beautiful dock and a beautiful boat, and I'm just laughing and crying at the same time.

"Money never mattered to Monty," Vally explains, "so he never thought it mattered to me. He wasn't cheap. He'd give you his money—if he had it. He just never had any."

Sunday morning, the team went to breakfast at Perkins— sometimes directly from the Magic Lantern, sometimes directly from a "date"—and told stories about the night before. A win on Saturday meant they deserved that Saturday night, and deserved

93

that Sunday morning, too—and having that Saturday night and Sunday morning meant that they couldn't wait for another tough week of practice ahead, in order to win on Friday and Saturday and have that Saturday night and Sunday morning again. "It seemed we could do this forever," says Vally.

Conditioning skates at the end of practice, extra circuits in the gym: when you lose, you need to do more; when you win, you can't do enough. If the energy dropped a bit, there were ephedrine pills—they weren't illegal at the time—and vials of ginkgo biloba. "We got the ginkgo from one of our teammates, who got it from his dad, who got it from some Japanese importer. It was our secret stuff, our super pill, and before a game we'd all stand around the garbage can in the middle of the dressing room, yell 'cheers,' and throw it back."

When you win you can do stupid stuff, and nothing seems stupid, and nobody cares. "During warm-up," Vally recalls, "we'd all have our helmets off, our hair would be flying as we went around the ice. We're sweating. We're looking at ourselves in the glass. We're cool. And we all had visions of playing pro hockey."

The Otters couldn't stop winning. They went 18–0–2 in their final twenty games and made the playoffs. In the first round, they played the London Knights, one of the league's best teams. On their roster were future NHLers Tom Kostopoulos, Rico Fata, Krys Barch, Chris Kelly, Alex Henry, and John Erskine; Jay Legault, Steve's friend from Peterborough, was the team's top scorer. The Knights won the first three games in the best-of-seven series; the last two crushingly in overtime. But then the Otters won, and won again. "The Knights thought we were done," says Vally, "and we just weren't." Erie won the sixth game at home in overtime, then lost the seventh in London.

Even though they had lost the series, Steve had never won like this before. He had never felt like he mattered so much.

The 1998 NHL draft was held on June 27 in Buffalo. With his December birthday, 1998 was Steve's first year of eligibility. The year before, Jay had been drafted by Anaheim in the third round, and Ryan Ready by Calgary in the fourth; a year before that, Colin Beardsmore was picked in the seventh round by Detroit, Chad Cavanagh in the ninth by Washington. Nick wasn't drafted. Vincent Lecavalier was the 1998 draft's first overall pick. Other prominent first-round selections that year were David Legwand, Brad Stuart, Manny Malhotra, Alex Tanguay, Robyn Regehr, Simon Gagné, and Scott Gomez. In later rounds, Mike Fisher, Mike Ribeiro, Brad Richards, and Brian Gionta were also chosen, and in the sixth round, Andrei Markov, Pavel Datsyuk, and Steve's North Bay teammate, Chris Neil. In total, 258 players were selected. Of them, only 29 would play more than 571 NHL games, the number of games that Steve played in his career. Steve went undrafted.

For Steve, it was another Peterborough summer teaching at the hockey school, runs at the Lift Lock, workouts in the gym, field parties at the Shack, and deep, deep conversations about who-knows-what with Keats and the rest of the guys; then it was back to Erie. Several of the Otters' veterans had "aged out" of junior hockey, including Vally and the team's captain, Colin Pepperall. The new veterans—among them Shane Nash, who was the new captain, and Steve—would need to fill their roles on the ice and off. The team also had a new coach, Paul Theriault. Theriault had won two OHL championships with the Oshawa Generals, and had gone to the Memorial Cup final each time. Sherry Bassin, the Otters' manager, had also been Theriault's boss in Oshawa. Now Theriault was behind the bench in Erie, having not coached junior hockey for ten years.

The Otters never got on track the whole 1998–99 season. The new veterans weren't leaders in the way the old ones had been; Steve wasn't a good enough player to be "the guy" that the team needed; and Theriault seemed not entirely happy to be there. The team scored about the same number of goals as they had the year before, but allowed many more against. They missed Vally. As for the bus rides, Steve, now a veteran, sat at the back in the Pit, chewing tobacco, playing euchre, laughing his laugh.

Erie made the playoffs and lost in five games to Guelph in the first round. Steve had improved during the year, but he was getting older. His junior eligibility was up, and he was beginning to seem like just another player. But he still wanted to play. He wasn't ready to move on to a Canadian college. That would mean playing without the dream; nobody made it from there.

The 1999 NHL draft was held on June 26 in Boston. The first overall pick was Patrik Štefan, and then came the Sedins, Daniel and Henrik. Tim Connolly, who had played in the Quebec tournament, was selected later in the first round, as was Martin Havlát. Ryan Miller was chosen in the fifth round, as was Steve's opponent with the Marlies and the Otters, Tom Kostopoulos; Henrik Zetterberg was the 210th overall pick. The 1999 draft class has been described as one of the NHL's worst ever: 272 players were selected; only 112 of them played even one game in the NHL; only 19 played more than Steve's 571 games. Again, Steve went undrafted.

Yet Steve would still get his chance. The Peterborough Petes had a spot on their roster for an overage player, so the following season, Steve returned home, this time to play.

———

In December, the Petes held their annual skate with the fans. For local kids, Peterborough's Memorial Centre is like Montreal's Bell Centre; the Petes are their Canadiens. The kids come to the arena that day, put on their skates, step out onto the ice, and then their pain begins. "I was so nervous," Adam Babcock recalls. A kid at the time, he stood, looked around the ice, looked around the building, and with nothing else to do, began to skate.

The moment was not much more comfortable for the Petes players. They were teenagers, and beneath the bravado most of them were shy and awkward, except around girls. A few minutes into his solo skate, Adam felt someone come up beside him, bend forward, and ask him how he was. It was Steve. They started to skate, Steve doing most of the talking, Adam nodding, then Adam talking a little bit himself. When it was time for Steve to skate with other kids, he stayed with Adam. "I went from feeling small to feeling like the biggest guy out there," Adam remembers. He was seven.

A few days later, somebody in the Petes office called the Babcocks and asked if they were interested in billeting a player. They had done it before but their kids, Adam and his sister Amanda, who was five, were now at an age where they saw everything and took in who-knows-what—and every billeted player, whether he was on the Petes or not, was a stranger. The Petes office had been inquiring about a place for Steve. Mary and Terry Babcock said yes.

The Montadors' house on the lake was only thirty minutes away, but with Steve's schedule, and with Peterborough's unpredictable and often sudden winter storms, the Petes thought it better that he live in town. Steve was also nineteen and had lived on his own for most of the past three years. He had become independent. It was easier this way.

If there wasn't a game, Steve ate supper with the Babcocks; or if practice kept him late, there were leftovers in the refrigerator for him to warm up when he got home. When he was around, which was much of the time, he did what the family did. He watched TV with Amanda, who was about a quarter his age. Amanda liked cartoons; Steve didn't. Steve learned to get to the TV first to control the remote. Amanda learned to get to the TV even earlier, turn on her cartoons, and hide the remote in the freezer. She knew he would never get up to change the channel.

On weekends, he would go bowling with the family, or push the kids on the swings. He played ball hockey with Adam on the family's driveway, and sometimes his teammates would come by and join him and Adam in a game they called "rebound hockey." Steve would stand in front of the net, a tennis ball on his stick, his teammates next to either post. They would put Adam in goal. Steve would shoot, and if there was a rebound, his teammates would pounce on it. "If they scored," Adam recalls, "it was a point for them. If I covered it up, a point for me. It was torture really, but Monty always kept a watch out for me." If he could see that Adam was getting frustrated, he would catch the eyes of his teammates and they would back off a little. "And if I stopped one," Adam recalls, Steve would shout like an NHL announcer: "Oh, and Babcock makes the save!"

Steve continued to visit the Babcocks every summer, even after the Montadors had relocated to Toronto and Steve had made it to the NHL. During his last season in Chicago, Steve invited Terry to go on a dads' trip (later these became moms' and special parent and friends' trips as well) with the Blackhawks. He had taken Paul on a similar trip a few years earlier. To a hockey parent (or friend), the only future more impossible to imagine than having a son make the NHL is the chance to share that future with him on the road; to live the life themselves. Terry travelled with Steve to Florida for games against the Panthers and the Lightning, then back to Chicago for a game at the United Center. Terry had known Steve for more than a decade by this time, but almost entirely in the environment of Peterborough, and within his own household. Now he'd be seeing Steve as a member of a big-time team that had won the Stanley Cup the previous year with a roster of big-time stars. How would Steve be?

"It was like he wanted to take me by the hand and show me his world," Terry says. "He wanted me to be a part of everything he was doing."

"He had this amazing gift for making you feel that you mattered to him," Mary recalls. "He would come back in the summer, he had all this money, it wasn't like he couldn't do whatever he wanted. But he hung out with us. He did it because he wanted to."

Adam is now twenty-three. His games of rebound hockey with Steve on the driveway turned him into a goalie. Recently, he got a new mask. On the top of it he has written "Monty"; on the bottom, "1979–2015."

The Petes were a mediocre team. But again, Steve was getting better. And again, he was getting older. When the 1999–2000 season ended, his junior career was over. He was twenty years old, and facing the same choice he had the year before. He could take the scholarship credits he had earned from his junior teams and go to a Canadian university, play a few years, keep improving, and hope. He had taken some courses at Mercyhurst University the second year he played in Erie. But he didn't want to go to college. He still had his hockey dream. Besides, going back to school would prove that he was wrong and Paul was right.

Calgary's general manager, Craig Button, had seen Steve play as he had scouted the OHL's stars the previous years. Button knew that there wasn't much difference between a fifth-round pick and someone like Steve. Neither is good enough at twenty to play in the NHL, and both need time to develop. They would both have to work harder than they had ever worked before in order to make it, and not everyone is willing to do that. Button liked Steve's spirit, and offered him a contract. Steve signed.

In February 2000, in a Petes game, at a moment that didn't seem to matter at the time and in a way he later wouldn't remember, Steve got his second concussion.

CHAPTER NINE

Dr. Karen Johnston realized early in her neurosurgical studies that, as she puts it, "you begin to understand better how the brain works by knowing how it doesn't work when it's broken." She began her career at the Montreal Neurological Institute (part of McGill University) around the time Steve suffered his second documented concussion, in 2000. Her focus was the "traumatized brain."

Brain trauma wasn't about concussion at the time. Concussion was a small, peripheral field dominated by observation and anecdote, not by science; it was driven by team doctors in all sports, most of whom were orthopedists because most sports injuries involved joints or bones. The doctors were beginning to see more clearly the effects that concussions had on their players, but because brain X-rays didn't disclose anything broken or torn, these athletes kept on playing—partly because, as players, why wouldn't they, and partly because, if there was no visible damage, their coaches assumed they would.

Johnston arrived in the field just as the "brain people," as she describes them, were becoming interested in concussion itself: in the pathology of it, and what it really means to the brain. Functional MRIs had been developed and were becoming more widely available—and these scans were able to show what had never been seen before. Neuropsychologists were getting more involved, too. Concussion had never been an invisible injury to them. They saw it every day in the symptoms and behaviours of their patients. They knew that if someone was doing weird and destructive things, something wasn't right.

Johnston treated some of the players from the CFL's Montreal Alouettes. She was not the team's doctor, but she watched their games from the sidelines and saw up close the effects of concussion and how the players' injuries occurred. She could also see the challenge of being a team doctor—on the one hand, there were the single-minded needs of the players and coaches, the passion of the crowd, and the urgency of the game as it ticked away; on the other, the doctors' profound obligation to do what is best for the patient, and their absolute inability in almost every case to know what was best, because of the mountain of things they did not and could not know. A player's dizziness, disorientation, headache— what did they signify? Were they of the sort we all have, ones that quickly go? Or something more? And what was especially difficult was that these team doctors had to decide *instantly*, as if they really did know.

Doctors who treated patients with concussions were looking for some certainty and guidance. If they weren't able to know all they needed to know, they wanted at least to be able to apply the best of what they did know according to some accepted standard. "Everybody was trying to hang on to something," Johnston recalls. The most obvious indicator—a player having been knocked out— wasn't good enough (it would eventually be dismissed as an indicator entirely). Doctors were seeing too many other symptoms that seemed to matter just as much. Therefore, a grading system was needed. One of the early leaders in this effort was Dr. Robert Cantu, a neurosurgeon in Boston, later well known for his work at Boston University.

Cantu's initial focus had been boxers. It was in boxing, many years earlier, that blows to the head in sports had first gained attention. In an article in the *Journal of the American Medical Association* in 1928 entitled "Punch Drunk," Dr. Harrison Martland expressed in scientific form what boxing observers had long noticed and what fans could see with their own eyes—fighters with slurred speech, tremors, and unsteadiness in their gait and balance. These boxers, who suffered repeated blows to the head, were described at the time as being "cuckoo," "goofy," "cutting paper dolls," or "slug nutty." Their condition came to be called "dementia pugilistica."

Around this time and for many years after, punch-drunk characters were often depicted in movies. They dressed in a funny way; they talked funny and said funny things. They were usually a sidekick to the star, like Sach in the Bowery Boys movies; they were the comedy relief. There was nothing sad about them. Their role was to chime in with the kind of wisdom that made everyone else, who looked and sounded smart, seem stupid—the way kids in sitcoms do. Red Skelton, a popular TV comedian of the late 1950s, played a recurring character called Cauliflower McPugg, who—with his cap on sideways, his head twitches, and his nasal, slurred speech—was hysterical. McPugg was a boxer.

In real life, most boxers were immigrants or African Americans, who were doing what immigrants and African Americans at the time had to do to get by. To many white Americans back then, they were seen as lowlifes who were just going to fight anyway—so why not in the ring to make a few bucks and maybe become a "some-body." And because they weren't like everyone else, when they walked around punch-drunk people would laugh at them and not feel sad.

Then along came Muhammad Ali. He was beautiful and smart, he did float like a butterfly, and he was definitely not a lowlife. It wasn't funny to see Ali in the years before his death. It's not funny to see football or hockey players now—local heroes who can't remember, and whose thoughts won't string together. It's hard to recall the last punch-drunk character in the movies or on TV. Punch-drunk isn't funny anymore.

But until recently, boxers seemed of another world to scientists as well. Boxers get knocked out. Football and hockey players rarely do. On an MRI, boxers' brains often look different; football and hockey players' brains look normal. In boxing, hits to the head are purposeful; in football and hockey, they are mostly incidental or accidental, less frequent, and are delivered with less force—except in hockey fights. That made the sports seem different, and led them to be thought of in different ways even by scientists. It took a long time for scientists to see that the issue was hits to the head, not about how they happened; and that boxing, football, and hockey were not disconnected worlds, but on the same hits-to-the-head continuum. For Johnston and medical colleagues engaged in other sports, it was this recognition that came to link their research and their work.

"We learned to connect the extreme end, which was boxing," Johnston says, "to people who were getting hit and experiencing symptoms but who were not being knocked out, and were not having abnormal imaging. We think of concussion now not as a structural injury to the brain—because when we do these scans, they are normal—but as a functional injury, as the brain not working in certain ways." If there is something functionally wrong, there must be something wrong whether they can see it or not. And once the scientists and doctors thought to make the connection between the sports, they began to think: Why would a fist in hockey be different than a fist in boxing? Why would an elbow or a shoulder be different than a fist?

This had been so obvious. But they couldn't see what they couldn't imagine, just as they couldn't understand what they weren't looking for. Neurosurgeons and MRIs couldn't see much of the damage present in a structural brain, but neuropsychologists could see that damage by looking at the *functional*, or dysfunctional, brain—by looking at people's symptoms and behaviours. It was seeing the structural through the functional.

In the early 2000s, more time and resources were being dedicated to the field of concussion. Studies were being carried out and articles were being written—published not just in scientific journals but in newspapers and magazines. Sports medicine conferences included sessions on concussions. With all this attention and interest, scientists and sports medicine people began to collaborate. Concussion was still a field on the margins of both sports and science—and sports medicine was still on the margins of medicine—but whereas earlier, when each specialty had been too small to fight big fights, they had fought smaller ones among themselves. Now, with bigger ones to win, they began to fight together over the profound impact of concussion itself.

One more thing proved immensely important. Work on concussions had been carried out by pockets of single-minded people involved in different sports on different continents. Though they had all decided that concussions mattered, most didn't know each other or know each other's work. But the more advanced science and higher public priority gave them greater reason to connect, and with the Internet they now had the means of doing so. The importance of instant communication cannot be underestimated, Johnston says. It allowed them to begin thinking about a global conference, the effect of which would be profound.

In 2001, the first International Conference on Concussion in Sport was held in Vienna, organized by the International Ice Hockey Federation (IIHF), the International Olympic Committee (IOC), and the Fédération Internationale de Football Association (FIFA). Johnston was chair of the Concussion in Sport Group that wrote the conference's final statement. Despite the fact the conference was held less than two months after 9/11, about 150 people from 15 sports organizations attended.

Grading systems were the conference's focus. "The Vienna meeting was the turning point," Johnston says, "because people declared their will to start working together, to start driving the science forward, not just the anecdotal stuff. That was a really big deal. At that meeting, we first started to hear about some of the great neuropsychological approaches. There were imaging studies. There was some good science that was just starting to take off. The other conferences that followed had the same goal in mind, but the first one changed the concussion world."

Two years later, in 2003, Johnston and other Canadian scientists decided it was time to take this new information on the road. They called it the "Concussion Roadshow." It was organized by ThinkFirst, the group headed by Dr. Charles Tator that had Paul Montador as a member of its board.

"We felt a necessity to take the message out to the public and to the teams," Johnston says, "because we were starting to learn what symptoms were important, but also because doctors still thought that loss of consciousness was the big thing, [and because] teams had no clue what to do, and players were going back out and playing when they were injured. And we were starting to know that there were way more concussions than we thought."

They travelled around Canada, gave talks, held conferences. "The presentations were built around athletes telling their story," Johnston says. "We had talks about the background science designed to be digestible to a wide audience. We talked about protection, about rehabilitation, about some of the signs—what we know and what we don't know. A team doctor talked about some of the issues that a team doctor faces. In the audience, we had athletes who had been concussed. There were medical doctors, physiotherapists, athletic therapists, coaches, and parents. And media." Many more people were becoming aware of the problem of concussions in sports. Johnston and the others were achieving what they had set out to do.

As scientists, their job was to study, learn, and apply what they'd learned. But with the roadshow, they had gone a step further. They had shared. Now the public knew that concussions were not just about "feeling woozy," "seeing stars," or "having your bell rung." They are brain injuries, and like injuries to a shoulder or a knee, they almost always go away—they heal—but sometimes they don't, or they don't heal completely. With a knee or a shoulder, sometimes function is affected, strength and mobility are lost or diminished, pain is increased; a shoulder that could once throw a ball ninety-five miles an hour can now throw it only eighty-eight. And injuries that do go away sometimes come back twenty or thirty years later, before regular advanced aging brings them back.

It's the same with the brain. Concussion is an *injury* to the *brain*. It can lead to the brain's diminished function, whether temporarily or for good. Losing seven miles an hour on a fastball is one thing; losing your equivalent capacity to solve problems or make decisions is quite another. Johnston and her colleagues had a message for the public: reducing the number of concussions is important.

In 2004, the second International Conference on Concussion in Sport was held in Prague. Two more followed in Zurich, in 2008 and in 2012. Another was held in Berlin in November 2016. The Mayo Clinic also held summits on concussions in hockey in 2010 and 2013. The scientists and sports medicine practitioners in the field were moving ahead. They were conducting more studies; they were using what they learned to better diagnose and treat their patients. The public were becoming more aware of the problem and of the best solutions. The decision-makers—coaches, on-ice or on-field officials, team owners, league and player executives—could then take this knowledge and apply it to their sports because now they knew, too. Because whatever uncertainties remained over the origins of CTE and other neurological disorders, over who gets them and how—through actions incidental or accidental, contrary to the rules or not—there was now no doubt, *none*, about the connection between blows to the head and brain injuries, between brain injuries and the resulting lousy things that happen to a person's life.

Johnston and the others had built the awareness; they knew the decision-makers would come.

But the decision-makers didn't come. At first, perhaps, they didn't know about the life-affecting impact of concussions, but then they did—and they ignored, then denied, then considered the possibility, then acknowledged the possibility but with caveats. Then they took some action, then emphasized how difficult it is to take action, then emphasized the actions they *had* taken, *especially* as compared to other sports, other leagues, other anything, *especially* considering all the scientific doubts that still remained. In doing so, they avoided the only real question: Is the dimension of the actions they have taken consistent with the dimension of the problem they face?

When the decision-makers were slow to respond, Johnston and others grew frustrated. They tried patience. They tried perspective—*we're a lot further ahead than we were ten years ago.* They tried understanding—*change takes time.* They tried generosity—*I think Gary Bettman really gets it.*

Busy with her patients and her research, Johnston occasionally still steps back, seeking perspective, and sees how slowly actions—true, meaningful actions—are being taken. She has decided it is her fault. She hasn't gotten the message across well enough, she thinks. "It has led me into the field of KT or 'knowledge translation,'" she says. "How do we take this information and make it digestible and accessible to various groups of people?" She offers an example: "The way an athlete learns information is not necessarily the same way a neuroscientist or a family doctor does. So what models of education do we need to develop, and deliver, so that we are not just handing some pamphlet to everybody and they're just trashing it? Shall we make another video? Shall we do this online? Does it need to be interactive? Shall we make an app for that?"

Johnston says she doesn't have an answer. "It feels like a desperate measure, trying to find ways to convince people to buy in. Ultimately, maybe we need marketing people doing this. I don't know. I'm a brain surgeon. What the heck am I doing trying to figure out ways to educate physiotherapists or hockey coaches about concussion? I was never trained in those skills. So, once again, you start collaborating with other people. All of this stuff is so outside the box from whatever I thought I would be doing."

Johnston tries to be optimistic. "It's why I like looking after my patients," she laughs. They do what she tells them to do—*mostly*. But she knows that's not good enough. She doesn't want to have to see the people she is treating. She can help many of them be better than they were when she first saw them, but she knows that they would be so much better off if they had never been injured, if the games they love to play and can't stop themselves playing were safer. But these decisions, she knows, are not hers to make.

Yet she sees hope. "Some schools have taken this on. Some teams have. Some of the stuff that's come out of the NFL studies, the problems with dementia and mood disorders and suicide. They create a forum and a voice to talk about this. The athletes coming forward; they are the best spokespeople." This is all about better "knowledge translation," in Johnston's words. But KT isn't always the problem. Others have to decide that they want to listen in the first place.

Decision-makers know who decision-makers are. They know the difference between influence and authority. They know that influence may reside in someone else's voice—someone like Johnston— but that authority resides with them. They know that they can say to Johnston or to everyone else—the scientists, researchers, media commentators, parents—"You and thousands and millions of others like you might be completely right. But you are not sitting in this chair; this is my decision not yours, and I have the right to do what I want to do." Parents know that coaches make decisions that they, as parents, cannot make themselves; coaches know the same about minor hockey officials, and minor hockey officials know that about leagues, and leagues about provincial or state associations, and provincial or state associations about Hockey Canada or USA Hockey.

And Hockey Canada and USA Hockey know, and the IIHF knows, that the NHL makes decisions that they cannot make. The NHL is the big decision-maker in hockey. It sets the tone, it determines the direction, because it creates the dream.

CHAPTER TEN

Gary Bettman never played the game. And he is American, a fact that has never gone unnoticed by Canadians. He attended Cornell University, where hockey is the biggest sport on campus, but hockey had not been a priority in his life before he became the NHL's commissioner. When he took on that role in 1993, it instantly became his job to run a league that dominates the direction of the sport, worldwide, for all the players and fans who absolutely love the game. Bettman has not always had the trust of those players and fans.

He was also taking over a league that did not function much like a league, which made the NHL just like every other major North American professional sports league. For a long time, these leagues had operated as little more than a collection of individual teams with very individual owners, who ran what they owned the way they wanted to run them. Their message to the league officials whom they employed was: "Schedule the games, hire the referees, and get out of our way." The teams made their money locally; it was certainly not the leagues that put "bums in seats," or company logos around arenas and stadiums. Leagues were understood by team owners as an expense—apparently necessary, but one that needed to be minimized.

This perception began to change with the National Football League in the 1960s and 1970s. The league's commissioner, Pete Rozelle, believed that a strong NFL would make a strong Chicago Bears or New York Giants even stronger, and he reinforced his case with team owners when he negotiated a big new network TV contract for the league.

David Stern believed the same of the National Basketball Association. Many of the NBA's franchises had been weak for decades when Stern was hired as the league's commissioner in 1984. He knew that the NBA's great stars and great teams gave presence—and financial value—to the rest of the players and teams, who were not great. Again, it was a big network TV contract that persuaded the owners of the league's value. Bettman had been third-in-command at the NBA before he was hired by the NHL, and had learned its lesson: a strong league is critical to the success of its teams. His problem, as it had been for Rozelle and Stern, was that for this to happen, he needed to sign a big TV contract in the U.S., which—given that hockey was nowhere near as popular as football, base-ball, and basketball—was far from certain.

Bettman had a shaky first decade. He did get his initial big TV contract in the U.S.—at least, big by NHL standards—but that was followed by one that was smaller. The league had decided even before Bettman was hired that, to be relevant in the U.S., the NHL needed a fully national U.S. footprint. It developed a "Southern Strategy," led and implemented by Bettman, which by expansion or by franchise transfer saw teams move into San Jose, Dallas, Anaheim,

Florida, Colorado, Phoenix, Carolina, Nashville, and Atlanta; later Columbus and Minnesota, were the only northern exceptions. The results were decidedly mixed.

Bettman also had some bad luck. The Canadian economy turned downward, Canada's federal and provincial governments ran up large deficits, commodity prices fell, and the value of the Canadian dollar plummeted from just over 80 cents to the U.S. dollar in February 1993 when Bettman began to 63 cents in August 1998, to its all-time low—61.79 cents—in January 2002. Most of the revenues for Canadian teams were in Canadian dollars; all player expenditures were in U.S. dollars.

Toronto was a rich enough city for the Leafs to cope with the currency gap and to compete with U.S. teams for free agents; Montreal and Vancouver less so. Calgary, Edmonton, and Ottawa had great difficulty. Their teams and their fans were losing confidence; they weren't sure when and how they could compete at the top again. All of this—combined with the rise of European and American players in the NHL, Canada's disappointing finishes in world and Olympic championships, and the departures of the Quebec Nordiques and Winnipeg Jets to the U.S. in the mid-1990s—made this American who had never played the game a very unpopular person in Canada.

Bettman wasn't much more popular in the U.S. New cities got their teams, but many of the teams couldn't seem to emerge from their apparently perpetual struggle to survive. It seemed that the commissioner's job, much of the time, was to keep the many franchise fires from getting out of control, then find new owners to stabilize things a little until the next fire popped up. With all this uncertainty, franchise values were not increasing the way they were in other sports, which didn't make team owners happy. And what made them even less happy were player salaries. Despite the fact that the NHL was a distant fourth in popularity among professional sports leagues in the U.S., its players, on average, were paid more than players in the NFL—and not much less than those in Major League Baseball. In fact, their salaries had increased by a larger percentage the previous decade than those of players in all of the other leagues except basketball. The NHL also was the only league, other than MLB, that had no salary cap. To NHL team owners, this wasn't the way things should be.

The existing collective bargaining agreement between the NHL and the National Hockey League Players' Association (NHLPA) was set to expire in 2004. The owners were looking for greater revenues, higher franchise values, and lower expenditures. They wanted a big network TV contract in the U.S., and the wage certainty that a salary cap would bring. It was Bettman's job to deliver this. These were off-ice issues that fundamentally affected the on-ice game: how it is played; the playing life and non-playing lifestyle of the player; and, most crucially, who controls the game.

Bettman is a businessman with a lawyer's training and instincts. This is who the league hired, it is what Bettman wanted to be, it is what the league got. In his new role as commissioner, he attended games and was visible in the seats. He watched; he listened; he learned. He learned enough to ask the right, probing, revealing questions. He weighed arguments; he applied logic. But he was not an expert in hockey. He hadn't been a player. He wasn't Canadian. He couldn't pretend he was any of these things, and he knew how ridiculous it would be to try. Sports fans can detect poseurs light years away.

To the owners he would be businessman and lawyer, and to the fans he would surround himself with "real hockey guys." Heart and soul ex-players, their pedigree displayed in stitches and limps earned in the trenches of the game. Canadians—maybe even those from small towns—who had lived the story of the game, who were authentic, who for Bettman would be the game's institutional memory and its conscience. Bettman could learn the language of hockey, but they knew its idioms. Anything about the game itself went through them. This only made sense; it was also publicly defensible. If ever Bettman got too lawyerly or said something cringeworthy, he had people around him to bring him back to the real world: To how things are done. To how hockey players think. To how the game is played.

Meanwhile, hockey continued to change. The Russians had arrived in full force. Not only as solo acts—Sergei Makarov in Calgary, Alexander Mogilny in Buffalo, Pavel Bure in Vancouver—but also, in Detroit, the "Russian Five"—Sergei Fedorov, Igor Larionov, Vyacheslav Kozlov, Vladimir Konstantinov, Viacheslav Fetisov. This was a whole five-man unit that had grown up with passing, where players without the puck were more important than the player with it, where greater and greater speed could be realized. There the future was on display on NHL ice surfaces for players, coaches, fans, and the media to see—giving players and coaches the opportunity to do the same and to experience all of its possibilities themselves. That regular season, the Red Wings won an NHL record sixty-two games.

The speed of the game was picking up. Shifts got even shorter—about forty-five seconds long—and the game got faster still. Even the language of hockey changed. Goal calls changed: "He shoots, he scores!" became "Shoots, scores!" or just "Scores!" There was no time to say more. Phrases like "ragging the puck" and "second wind" all but disappeared. With so many players chasing the puck carrier, there was no time to "rag" it, and players, moving at a sprint, had to get off the ice before their first wind expired. A new phrase appeared in the mainstream hockey lexicon: "finishing your check." After a puck carrier makes a pass, a checker continues towards him and runs into him, finishing the check. The phrase didn't even exist in the 1950s—nor even during much of the 1960s. The players, still playing two-minute shifts, coasting, circling, weren't close enough to each other to finish their checks.

And one telltale story from the 1940s: of the great Bill Durnan, hockey's first and only ambidextrous goalie. On a play coming at him from an angle, he liked to hold his stick on the short side, the side nearest to the shooter, to enable him to catch shots with his other hand to the long side. The game moved slowly enough that when the puck was passed from side to side, he had time to switch his stick from hand to hand.

Fighters also changed. The heavyweights arrived—Bob Probert, Mike Peluso, Stu Grimson, Gino Odjick, Todd Ewen, Donald Brashear, Krzysztof Oliwa, Georges Laraque. Even fighters shorter in stature—Tie Domi, Rob Ray, Kelly Chase, Paul Laus (who had thirty-nine fights in the 1996–97 season)—were heavyweights. Most still tried to contribute fully as players, to get more ice time, to help their teammates in other ways, and after practice and in the summer they worked to build up their skills. But they also trained to become better fighters. By the 1990s, it was hard to pretend anymore that fights were the inevitable result of a game that moved superhumanly fast inside an enclosed ice surface, where collisions were unavoidable, where anger resulted, where wrongs had to be righted, where a fight was a release of potentially dangerous pent-up emotion.

For why were these fighters so angry? Because their team was behind and they needed to go out and "change the temperature of the game"? Because their talented teammates were angry at being hit, and as fighters they were angry on their behalf? These fighters weren't on the ice long enough to be angry for any other reason. Fighting wasn't a release of emotion for them—it was exactly the opposite. It was a fighter's chance to *add* emotion to the game. But since fewer fighters fought more fights, since it was becoming undeniable their role was to be a goon, their role would need to be romanticized. If you weren't fighting to right your own wrongs, you had to be fighting for your team, for your teammates. Literally. Risking life and limb *for them*. Literally. Fighters now mostly fought those who had committed the wrong to someone else, or fought other fighters. There was a code. If fighting wasn't about anger, it had to be about honour.

Fighters of the 1950s and 1960s couldn't lose a fight, because then they would have to prove themselves again and again, and fight more often. If they won routinely and devastatingly on the rare occasions they fought, they earned a reputation and others stayed away. Nobody, after all, wanted to mess with Gordie Howe. In the 1970s, the Flyers changed that. Dave Schultz didn't win all of his fights, he just kept coming back at his opponents until they weren't willing to fight anymore. It was the same with Bob Kelly, Don Saleski, Jack McIlhargey, and whoever the newest and toughest on Philadelphia's roster was. But by the 1990s, a fighter had to win every fight again. There weren't many fighters, and they were all very good at what they did. They could do serious damage, and be seriously hurt. It was gunfighter against gunfighter: a showdown. The fights weren't emotional; they were circumstantial—if your team was behind by two goals but there was still time for a comeback, you fought; if your team was behind by more than two goals and the game was lost, you fought; if a star teammate was hit and that wrong needed to be righted, you fought—this was about pride, justice, or changing the tone of the game. For a fighter, if his team is down two goals and he is put on the ice, what is he going to do to help? Score two goals? Fighters fight and dancers dance. By the 1990s, fighters couldn't lose, because winning fights was what they did.

For the fighter, *everything* continued to escalate too. He played less often; he had to make his mark faster and with more certainty; he had only one way to keep his place on the team. He had to fight guys as big and tough as he was. Six-foot-four, 230 pounds. Bare knuckles. Fighting hurt more than before. No matter the injury or fear or pain, a fighter had to be ready. He had twenty seconds to prove himself. Painkillers and anti-anxiety medications became parts of his life.

The names of former player-fighters who showed symptoms of brain injury, sometimes before, sometimes after retiring have started to appear—Bob Probert, Wade Belak, Derek Boogaard, Mike Peluso, Marc Potvin, Stu Grimson, Gino Odjick, Todd Ewen. Fewer concussions were reported in earlier decades. There were fewer doctors around a team to diagnose them. Players and coaches were less aware. Players continued to play because that's what players do. But with the game moving more slowly in those days, with fewer collisions and collisions of less force, with fewer fights— maybe fewer concussions had been reported because there *were* fewer concussions.

Bettman wasn't a hockey guy, so he depended on his hockey guys—who saw the much faster, more exciting, more dangerous present through the prism of a very different past. Through a past of legend and lore; through their own memories, like generals who fight the next war using the last war's strategies, employing war horses against machine guns. Bettman had grown more confident about his upcoming collective bargaining agreement (CBA) showdown in 2004 with the players over a salary cap; but he was no more confident about the *feel* he needed for the game on the ice. When Bettman is not sure of himself, he retreats into his lawyer's self. He listens to his critics to pick holes in their arguments; to create doubt, to keep them at bay, to avoid the larger issue about which he is less certain. A salary cap was where he should put his attention, he knew. It was the one big issue that would determine so many others—for the future operations of the game and for himself.

Other aspects of the game were becoming crucial, too. Players always give *everything*, and they will give whatever the *everything* is at any moment. Commitment, time, training off-ice and off-season, more hours, more months, shorter shifts, skating faster, colliding harder and more often, fighting more punishingly, blocking more shots, playing with more injury and more pain—whatever it takes. By the early 2000s, *everything* was a lot more than it used to be.

It was in this ever-escalating game, in this NHL, that Steve would play his career.

CHAPTER ELEVEN

It was September 2000. Steve packed his gear, piled it into the back of his new pickup truck, and hit the road for Saint John, the home of Calgary's AHL affiliate. The kid who had almost always been the youngest, who had never been the best, who had gone undrafted, was now a professional hockey player at twenty years old.

He spent the first night of the journey in Montreal. It was at Paul's suggestion, and Steve had agreed. The next morning, Steve met with Gisele Bourgeois, a psychologist and performance coach who had worked with some senior executives at Johnson & Johnson. Steve and Gisele also met that evening and the morning after, for two or more hours each time.

Gisele was different from Steve's other coaches. She didn't tell him what to do, nor think of him in the context of a team. She wanted him to think of himself as being at his own centre. She wanted to help him understand what *he* wanted, what *he* needed, what *he* hoped for. He had played hockey since he was three. Hockey had filled most of his days and weeks, most of his doing and dreaming, year after year since then. He was good at hockey. But did he play hockey because he loved it, Gisele wanted to know, or because hockey was just what he did? She wanted to hear Steve answer questions like this because she wanted him to hear his own answers. She wanted him to reinforce the certainty he already felt, that he had demonstrated through his actions all of his life, because what he was about to undertake and wanted to achieve would require an immense commitment. She wanted him to know—and for him to say out loud: "I am at the wheel driving to Saint John to play for the Flames because *I* want to. This is *my* choice. This is *my* life."

Steve was excited and fearful in their first session, Gisele recalls. Her first impression was of "someone who at a young age had made up his mind that this is what he wanted to do, and who, when he makes a decision, is all in." He had fears, but they weren't about living up to the expectations of others or about living up to past successes. They were about what was ahead. He wasn't a superstar. He had never been able to shape his environment to his own wants and personality. He'd had to adapt to every new team he played on. *How will I fit in now?* he wondered. *What will my new coach be like? Can I do it?*

Gisele noticed that Steve was very aware of his impact on others, especially his teammates. He wanted to make other people happy. This wasn't to get validation from others, she believed, but rather a simple "why wouldn't I?" Isn't that how any team, how any family, works? Your purpose is them; their purpose is you.

Steve had played in front of lots of people in his life, and had been written about and treated as special. But "I don't think he ever felt special," Gisele says. "He had an ego like every young man, [but] he knew that he wasn't the best player on the ice. He wasn't the most talented, and every time he was on the ice he knew he had to earn the right to be out there again. But he was okay with being one of the supporting cast. He knew that was important. He knew that he could make that significant, and he knew that early on. It gave him a certain humility, and kept him always searching for new ways to contribute to the team."

"It was maybe a quarter to twelve," Jim Playfair recalls, "just before noon, and Monty walks in." Playfair was then the head coach of the Saint John Flames. That night, October 13, 2000, the Flames were playing their opening game at home. The morning practice was over; Playfair had posted the team's starting lineup for the game. Steve, who had made the roster as the seventh defenceman—managing to avoid being sent to their East Coast Hockey League (ECHL) affiliate in Johnstown, Pennsylvania—would not dress. "I had this really small office in the rink," Playfair says. "Monty sits down and says, 'I think you are making a mistake.' I said, 'About what?' He said, 'The lineup. I think I'm better than a couple of the guys that you've got playing tonight.' I said, 'You do?' He said, 'Nothing against anybody, and I realize you're probably playing them because they were drafted and I'm a free agent, but I think I can bring more to our team.'" Playfair thought for a moment. "I said to him, 'Why don't you do this—go home and get a bite to eat, and come back tonight for the warm-up.'" Steve left Playfair's office. "Ronnie Wilson, my assistant coach, looks at me and says, 'I'd send him down to Johnstown. Who does he think he is, coming in here?' I said, 'Tell the trainer to get him a jersey, because anybody who has the courage to walk into a coach's office and tell him he's made a mistake deserves a chance to play.' After the warm-up that night, I decided to put him in the lineup."

To Playfair, that moment defined Steve. The self-confidence, the trust he had in himself, his willingness to put himself on the line. Steve had told him how to fix his mistake before he'd even made it. And he did it without bad-mouthing anybody. "He made it seem that it wasn't about him, that it was about the team, because he thought it was, because it was," says Playfair.

"That was an important moment for Monty, but it was for me, too," says Playfair. "I think I became a better coach when I became a father." He had three young sons at the time. "As parents, we all want our kids to be more self-confident, more trusting of themselves, more committed to whatever strengths they have. We strive to instill these qualities in them. As coaches, we want the same in our players. A team is about the players. It's not about my ego, about me saying, 'I made the decision. I'm right.' If I listen to the players, their voices get stronger. *They* get stronger. Maybe that was one of my finest coaching moments, putting Monty in. I learned something that day."

Maybe Gisele had an impact sooner than she ever imagined.

Before the season began, the players and coaches each filled out "goal setting" forms—the players about themselves, the coaches about each player. Under the category of "major strengths," Steve wrote: "my competitiveness and determination; my skating and passing." The coaches wrote: "physical play, energy, competes; puck movement." As things he needed to work on, Steve mentioned "strong defensive play; one-on-ones; gap control; physical play." The coaches wrote: "defender who competes hard consistently."

"We had a different group of players that season," Playfair says. Three-quarters of them were in their third professional season. Their entry-level contracts were expiring; they were trying to earn new ones and had something to prove. The Flames had two former first-round picks, Rico Fata and Daniel Tkaczuk, who had played minor hockey with Steve in Toronto; some mid-round picks; and some free agents, like Steve. Not long before, these draft picks had been judged by the managers and scouts of the Calgary Flames to be among the best players in the world for their age. Yet at the moment they were chosen, none of them, except for a few prodigies, were good enough to make the NHL. First-round pick, ninth-round pick, undrafted free agent—it is the player who improves the most *from that moment* who will make it. It still matters to these players and coaches who was drafted and who wasn't—but it won't for much longer. Their draft status will be their identity until something else is. The draft pick who doesn't live up to expectations will carry around with him a stigma of disappointment that rarely goes away. The undrafted player who makes it will be surrounded by admiration. He will be the guy who worked and made himself into a player. Early in the season, Playfair had a message for his players: "For all of you, drafted or just signed, what being here really means is that you have now committed yourself to work harder than you have ever worked in your life. Because now there is another level you have to get to."

Of all the players on the team, Daniel Tkaczuk knew Steve best. They had grown up in neighbouring parts of Mississauga, had played on the Marlies, and had gone to the Quebec peewee tournament together. They had both moved to the Mississauga Senators, beaten the Marlies, and won the GTHL championship together. Their mothers, who had driven them to the rink, watched them play, and feared and hoped together, were good friends. Daniel and Steve had also played against each other for three years in the OHL; when Steve stayed on for an overage year in Peterborough, Daniel had begun his pro career in Saint John, co-leading the Flames in scoring. Now their hockey lives had intersected again.

Seven years had passed since that season together with the Marlies, but not much was different, Daniel recalls. Steve was still a high-energy guy, charismatic, hard-working, loud, popular, always having to do something; some nights he was the best player on the ice, some nights the wheels of his wagon were falling off. But "his effort was always there," Daniel stresses. Daniel was still quiet, shy, serious, an observer, a thinker; "a hockey sense guy," as he describes himself, someone who aimed before he fired, when sometimes Steve did not.

Daniel remembers Steve with some surprise and a lot of admiration. Steve always had to "play through stuff," he recalls. "He was never seen as a top defenceman in our age group. He was never considered for Team Ontario or for the world junior team. Even to play AAA hockey, to play for the Marlies and the Senators, at times seemed a stretch." He was never not on the margins. "Playing through stuff" means playing through whatever circumstance arises, team or personal—whenever it arises—and doing whatever you're asked to do. "You know how hard you've worked just to get to that point," Daniel explains, "to be so close, that you're going to push it no matter what." On every team he played for, Steve was looked to for "the physical element," as Daniel puts it. "You can't be soft. You can't say, 'I can't take any more of these hits. I'm not going to fight. It just doesn't feel good anymore.'" Your teammates depend on you. Your coaches and managers trust you. They believe in you—that's why they signed you. And you know that there's not much difference between you and the guy who is the seventh defenceman and not on the ice—or the guys who are eighth, ninth, and tenth, and not playing in the league. The biggest difference may be simply that you are physically able to play and they're not. Or they are able to play and you're not. And when you are asked to play, you play because you don't know when, or if, you're going to be asked again. Very few players now know the story of Wally Pipp, but every player knows the lesson of Wally Pipp in their bones. As the myth goes, mostly apocryphal, Pipp was the first baseman for the New York Yankees in 1925 when, because of a headache, he was given the day off. The next time his replacement, Lou Gehrig, was out of the lineup was fourteen years and 2,130 games later. By that time, Pipp was forty-six years old and living in Grand Rapids, Michigan.

Daniel remembers the final game of a tournament in Kitchener when he and Steve were playing AAA with the Mississauga Senators. Their opponent was the Elgin-Middlesex Chiefs, and future NHLers Joe Thornton, who was already six-foot-three, Mike Van Ryn, and Brian Campbell. Playing against Thornton, Steve had a choice. Most players, if not intimidated by him, became passive. They backed off, avoided the one-on-one contest that by size and skill they would surely lose, and waited for help. Steve wasn't as big as Thornton, but he was big enough, and he could skate. He was competitive; he could focus; he could live with humiliation and defeat. He went right at Thornton. "He got in his grill," Daniel remembers, and he stayed there, giving him little room to breathe. Thornton wasn't used to that, and the Senators won. Steve ended up on the tournament all-star team.

Early in a season, a team doesn't know what it has. The Calgary Flames had missed the playoffs the previous four years and hadn't started the 2000-01 season well. Players were being moved into the lineup, and out. Some were sent to the minors.

Saint John was also sorting out its roster. Steve played some games and didn't play others. Playfair worked with the defence in practice drills to get the puck out of their zone. "Passing from deep in your corner to the middle of the ice, that's insane," Playfair says.

But sometimes that's where the open ice is. As a defenceman, you need to read when to make that pass, and when not to—and be able to make that decision instantly. "One game," Playfair recalls, "Monty passes the puck to a teammate in the middle, it goes off his skate, into the slot, one of their guys takes a slapshot; it hits the crossbar and goes up into the stands. When Monty comes off, I say to him, 'That's *exactly* the play we want. Next time, trust yourself and make it again.' Monty had this 'but it hit the crossbar' look on his face. I said to him, 'They didn't score.'" It wouldn't be the last time Steve made a pass up the middle in his own zone, but the puck didn't always ring off the crossbar.

A player needs to know himself, and a coach needs to know his player, what he's good at and what he isn't, what he aspires to be and what his capabilities are. It's the only way their relationship can work. In December, the Saint John players and coaches were given new forms to fill out. The players were asked to rate themselves, and the coaches rated each player—excellent, good, adequate, or poor—in the following categories: pre-game preparation, pre-practice preparation, determination/battle level, work ethic during games, and work ethic during practices. Steve answered, respectively, *good, excellent, good, good,* and *excellent.* The coaches rated him *good, adequate, adequate, good,* and *good.*

Of the areas of his game about which he felt most satisfied, Steve mentioned his "work ethic; commitment." His coaches wrote about him: "willingness to compete (toughness); good skater and strong passer; his desire to learn pro game."

About those aspects of his game that he believed helped the team the most, Steve answered: "aggressiveness; high intensity with moving pucks; being physical." His coaches wrote: "playing physical game; plays within himself."

Calgary GM Craig Button reviewed these forms with his Saint John coaches, and kept them, so that one day if a player had problems and Button was looking for answers, he might find some clues here. Button still had these forms many years later.

By December, Saint John was in the middle of the pack in the AHL, at times doing well enough to make the players and coaches think that greater success was possible, but then doing poorly enough to make them wonder. There was a team-making moment just before Christmas. Wayne Fleming, coach of the Canadian team at the Spengler Cup tournament in Davos, Switzerland, phoned Button and asked for four of the Saint John players; Steve was one of them. Playfair called the four players into his office and told them the news. Then he and Ron Wilson met with the rest of the roster. Playfair recalls telling them, "This is a great opportunity for us. Are we a good team, or do we just have some good players?" Two or three of these players had been sitting out most games; now they got to play. The others had to fill bigger roles. "When those guys were in Switzerland," Playfair remembers with a smile, "we won all three of our games."

The players who stayed behind in Saint John proved to themselves, to each other, and to their Spengler Cup–playing teammates, that they were good, too. Steve and the others who went to Switzerland, tested in the playoff-like atmosphere of a tournament, came back better players. When they rejoined their teammates, they also realized that they weren't indispensable. "The second half of the season," Playfair recalls, "the players believed we had something special.

"Between those high moments," Playfair now says, "there was some misery. Some days when those kids didn't have a passion for this life anymore and thought I was holding them back." Steve's bad moments came, as they usually did, when he tried to do too much.

Playfair worked with him patiently. "First of all, with Monty," he says, "you had to understand he was always prepared. He worked in practice, and in the gym. He was always into the game. He didn't do things for his own glory. If the team needed something, he tried to give it. So his weren't lazy mistakes. They were winning mistakes. He was trying to make the winning play. Lots of players just want to get through their shift. Monty wanted to *win* every shift. Lots of players, when they make a mistake, they slam their stick on the ice, or bang the glass. They want you to know that they care. I don't remember Monty ever doing that. He knew I knew that he cared."

When the team had injuries to its forwards, sometimes Playfair asked Steve to play wing. "Some games I know he didn't want to do it, but it gave us our best chance to win. I told him that there would come a day sometime when a team, finalizing its roster, would have to make a choice between two players as their seventh defenceman: one who could give the team some flexibility and also play up front, and one who couldn't. I'm sure Monty thought that I was just giving him the company line." But Steve did it. And years later, when he was playing with Florida, Jacques Martin, the Panthers' coach and general manager, had to make a roster decision between Steve and another defenceman. "I opened the paper this one morning," Playfair remembers, "and there is Jacques saying something like, 'We like Steve because we can also put him up on the wing.' So I sent Monty a text: 'All those years ago that you didn't want to be a winger?' And he texts back: 'Yeah, that just made me a million dollars.'"

Playfair and Daniel Tkaczuk both remember a particular game later that season. It was on the road against the Providence Bruins, who had won the AHL's Calder Cup two years before. "They had some physical guys," Daniel recalls. "It was a crash and bang game, and the crowd was really into it." Playfair picks up the story: "They were running at us and pushing us around. Monty was sitting on the bench waiting for his shift and he leans back and says to me, 'Put me on.' He goes out, grabs somebody, and fights him." It was Lee Goren, a big, physical winger. Later, they fought again. "It changed the whole temperature of the game," Playfair says. "Monty just said to the other team, 'Enough. You're not allowed to do that.' He wasn't going to let us be on the receiving end any longer." The Flames won, 6–2.

After that game, Daniel recalls, the coaching staff and the players saw Steve differently. He had "stood up." He had given them a reason to see to the other non-physical aspects of his game. His occasional lapses seemed to matter less. He stayed in the lineup the rest of the year.

The Flames finished the regular season behind only Worcester and Rochester, with the third highest number of points in the league. They swept Portland in the first round, then beat Quebec, then Providence (who had upset Worcester in the second round), and advanced to the final against the Wilkes-Barre/Scranton Penguins, Pittsburgh's farm team. The Penguins won the first game in Saint John; the Flames won the next two; the teams split the two after that. The Flames headed home to New Brunswick for the sixth game, and a chance to win the team's first Calder Cup championship.

With one minute to go, Saint John was ahead, 1–0. This is a player's time, Playfair says; it is the players, not the coaches, who will decide the outcome. There was a time-out. The Flames captain, Marty Murray, had been with the team three years earlier, when the Flames lost in the finals to the Philadelphia Phantoms. Playfair had told Murray at the beginning of this year's playoffs that, if the team had a chance to win, he would have him on the ice at the end. During the time-out, Playfair gave the players their assignments and reminded them of a few things; but his real purpose, in the biggest minute of the biggest game of the season, was to get the players on the ice that he could count on most. Murray skated out for the faceoff. So did Steve.

Steve's brother Chris tells this story with a catch in his voice, because Steve had a catch in his voice when he told it to him. Fourteen years later, before Steve's funeral, Steve's mother Donna told Jim Playfair this story, because Steve had told it to her. The kid who had never been drafted, who needed to talk his way into the home opening night lineup, who the whole year had done the kind of mucky stuff that his team had needed even if he wasn't very good at doing it—*he* was sent onto the ice to bring the game home.

A minute later, the season was over. The Flames had won.

Lindsay has a full-page photo from *The Hockey News* hanging in her basement. With a jammed arena of heads as the background, it shows her brother mostly in profile, a baseball cap on backwards, his arms upraised, his eyes on fire, his mouth open in a roar of triumph, the Calder Cup in his hands.

When the next season began, Steve was back in Saint John. At 2:30 in the morning on November 23, 2001, Steve was awakened by a phone call. It was Jim Playfair. Calgary defenceman Igor Kravchuk had been injured, and Steve needed to get to Buffalo right away. He would play his first NHL game that night.

He got an early-morning flight to Toronto, but the plane got rerouted to London, Ontario, because of fog. The plane stayed at the gate in London, and passengers were instructed to remain on-board as the pilots waited for the weather to clear. When it didn't clear, the flight attendants eventually gave in to Steve's pleading—"I've got to get to my first NHL game!"—and allowed him to collect his bag and sticks from the hold and get off the plane. He rushed and got a taxi to take him directly to Buffalo. Paul, Donna, Chris, and Lindsay all drove down from Toronto to be there.

No goals were scored in the first period; Buffalo scored three by the mid-point of the second. Rhett Warrener was a defenceman for the Sabres at the time; two years later, he and Steve would be teammates in Calgary. "I can remember looking at the stats sheet that night before the game," Warrener says, "and thinking, 'Here's a kid who's played the last two years in the minors and has just been called up. Guaranteed he's going to want to fight and make a name for himself. I know that's what I'd do. I looked around our dressing room and I'm thinking, 'He doesn't want to fight Rob Ray [the Sabres enforcer]. I'm probably next on the list. I'm going to have to chuck the knuckles with this kid.'"

Instead, four minutes after Buffalo's third goal, Steve got into a fight with Denis Hamel.

In the third period, with the score at 4–0, Steve got an assist on a goal by a former Saint John teammate, Steve Begin. The Flames lost, 5–2. Steve had two shots on net and played just under fifteen minutes. "[He] was pretty good," Calgary coach Greg Gilbert said after the game. "He was pretty steady. That's what Steve Montador is going to be, a defenceman who moves the puck well and doesn't try to do too much with it." Earlier in the day, a Saint John journalist had asked Jim Playfair why Steve was the one called up. "We didn't say to ourselves, 'Well, he's the hot commodity of the day. Let's send him up there,'" Playfair said. "He's undergone steady growth and development. He's a very steady, strong player who moves the puck well, gets up the ice well, and gives us a physical presence. With all those factors, he'd earned the right to go and play in the NHL." Two days later, Steve played again, in Columbus, then was sent back to Saint John when Kravchuk returned from his injury. He would play a total of eleven games for Calgary that season.

Not long after his return to Saint John, Steve and his teammates and coaches filled out their December review forms. Under the categories of pre-game preparation, pre-practice preparation, determination/battle level, work ethic during games, and work ethic during practices, Steve rated himself *excellent, good, excellent, excellent,* and *excellent.* His coaches rated him *excellent* across the board. As to the areas of his game he felt good about, Steve wrote: "defensive zone play; understanding on-ice awareness/ice management; being confident about my game." His coaches wrote: "overcame a rough start; confidence—wants to be an impact player; physical presence—becoming a feared defender." To the question of how the organization could make him a better player, Steve responded: "expand the workout area." The coaches wrote: "time and patience." For the first time, Steve's coaches saw more in him than he saw in himself.

In his rookie year in Saint John, Steve had learned to survive. He had learned to be a regular. He had learned how to win again. Over the next few years, in Saint John and in Calgary, in the words of Marty Gélinas, "he learned how to be a pro." He had known how to prepare himself for games and practices; he knew the importance of the little details; he knew he needed to commit and deliver every night, and every moment of every night. But he got better at all of these things. Steve was young when he began his second season in Saint John—just twenty-one. Defencemen and goalies, hockey wisdom says, take longer to develop. Not because of the physical skills required, but because of the consistency and discipline that are needed to play such unforgiving positions—a consistency and discipline that maturity brings. Forwards have to create, and they fail almost all of the time because of the difficulty of what they are asked to do. It's to be expected. Not so for defencemen and goalies. Very few goals are scored in a game; very few can be allowed. Offence is about opportunities; defence is about mistakes.

Steve played solidly for the rest of the 2001–02 season, and the following year, when he was mostly in Calgary. He was getting better, but he wasn't yet ready to be a regular. Being ready is not something you wait for. It is something you prepare for. Steve was preparing himself for a season and a moment he couldn't be sure would ever happen. The Cup run of 2004 was still ahead.

In Calgary, near the end of the 2002–03 season, Steve was struck in the face by a puck and received several stitches. His injury was described as "facial lacerations." It was likely his third concussion.

CHAPTER TWELVE

The Cup run runneth over.

It was late June 2004. The bars along the Red Mile were still red with Flames jerseys; even a little green, with all the hard hats. The frenzy of Calgary's miracle run to the final had died down, but the glow remained. The city felt good: summer-good; playoff-good. Iginla and Kiprusoff were still the team's stars, but all of them were now heroes, Steve included. And with time finally to fathom what he had achieved, to the fans Marty Gélinas was even more than "the Eliminator." Three series-winning goals, two of them in over-time. How was that possible? Three playoff series wins; three wins over division champions; the Flames scoring just two goals in the last two games of the quarter-finals against Detroit, winning them both, one of them in overtime. How was any of it possible? In a few short days, their playoff run had gone from unbelievably exciting to legendary.

When a player wins, he celebrates as long as he can; until somebody tells him to go home. He squeezes every bit of pleasure out of the victory, because he doesn't know if that feeling will ever come again. The Flames didn't win in 2004, but it seemed as if they had won. The team partied like they had won. And Steve, with no regular girlfriend, his family more than 2,000 kilometres away, had no one to tell him to go home.

"Marty [Gélinas] and I used to train at the gym at the Father David Bauer Arena," Hayley Wickenheiser recalls. "That's how I met Monty." It was not long after the 2003–04 season had ended. Wickenheiser was twenty-five, a year older than Steve, and the best women's hockey player in the world. She had won her first Olympic gold medal two years earlier in Salt Lake City, and would go on to win three more. Calgary, the base for the Canadian women's team, had become Wickenheiser's home. She also had a little boy, Noah, age four, whom she had adopted—the son of her former boyfriend. "Monty would come and train at the gym. We became friends and hung out occasionally. He became like a brother, and was very kind to my son Noah.

"Monty was always a larger than life, happy-go-lucky guy. He would come into the gym and say hi to everyone. He was an NHL player, but he didn't take himself too seriously. He didn't think he was better than anyone else. This wasn't just an elite gym. I remember this one older gentleman, he had an oxygen tank and he was there for rehab, and Monty would always ride the bike beside him. The guy was probably eighty years old, and they'd just be chatting away."

Steve was also a confusing set of contradictions, Wickenheiser says. "He would come and go. He was pretty aloof when he wanted to be, and a very cerebral person. He wanted to talk about life, about all kinds of things in life. Publicly and outwardly he was this big hulking strong man, a hero in the run for the Cup; in reality, he was always trying to figure out who he was. What his place was. I'm not sure he ever felt he was worthy of what he had.

"It's a really insecure life, especially as a 5–6 defenceman. You don't know if you're going to be in the lineup, or even have a job, and he didn't have a lot of stability outside hockey in his personal or family life. And one thing about Monty, he was like a very open book. He would tell you everything. Sometimes you didn't want to know as much as he was telling you.

"Around that time," she recalls, "he started to spiral into cocaine and a few other things."

As a young teenager in Mississauga, Steve drank a little beer because most kids did. It was curiosity and rebellion. In North Bay, away from home, he drank a lot more, in that barber shop/hair salon getaway the players had. There were also the road trips with the team. A game is over, you're tired, but you're so pumped up you can't sleep. You just shared this intense two-and-a-half-hour experience with your friends and you want to keep sharing it. What are you going to do in the middle of the night? Where are you going to go? To your room and study? You're bigshots, kings of the world, and the beer only makes you feel bigger.

In Erie, Steve's drinking got worse—and for the first time in his life it had gotten in the way of his role as a player. He had been just a kid in North Bay; whatever he gave the team was a bonus. But in Erie, especially in his second year there, he was a veteran. The Otters needed him to be a leader on the ice and off. As a rookie, you just need to be one of the guys. When you're older, to lead, you need to be that and more. Too much beer made Steve seem too much like everybody else.

And during summers in Peterborough there was the Shack, where everybody brought along a two-four. Things would get crazy at times, but not too often. There were the downtown fights, but growing up in Peterborough, if you were a kid, things like that happened. Drinking was a chance to loosen up a bit, laugh too loud, and say what you'd never say sober. It was about having fun. And, as Steve's friend Mike Keating put it, "it gave you a little liquid courage to pick up girls."

While being a jock offered Steve and his Peterborough buddies more opportunities to drink, it also held them back. People knew them. If they saw them a little drunk around town it might make them seem like good guys. More than a little drunk, and people would talk. Word would get back to their parents, to their hockey coaches. Some of the Peterborough guys played lacrosse in the summer for the local team. The people who saw them drunk one night might see them playing at the arena the next. The team might lose. So they had to pick their spots. If there was a two-day break between games—hockey in the winter; lacrosse in the summer—they could drink, but as Keats said, "You didn't want to be playing guilty."

Throughout his years in junior, his two seasons in Saint John, even in his first year in Calgary, Steve seemed no different from any of his buddies at home. They were all college-aged guys, and whether in college like Nick, or working a regular job like Keats, or playing pro like Steve or Jay, they were living a college life. But things were beginning to escalate for Steve. He now had money of his own. He was now one of the young guys on a team again, and being one of the guys in the early 2000s, for some, meant not only beer but cocaine. Alcohol at night to loosen you up and bring you down; a few hours later, cocaine to give you the boost you needed for the next day's work. And Steve's work was every day, weekends included. He wasn't a regular on the Flames; he got almost no days off. He had a game or a practice almost every day—trying to keep his spot on the team, he had to be up for every practice, not just every game, and practices were at eleven. He didn't have the whole day to rest and recover, to get ready for the game that night as the regulars did. Steve was, as Wickenheiser says, headed into a "spiral."

One day at the gym, Steve told Wickenheiser his routine that summer: go out all night, sleep an hour or two, be at the gym at seven, work out, sleep in the afternoon, go out all night again. "I vividly remember one day he came into the gym and had two mismatched socks on, a black one and a white one, and then proceeded to throw three hundred pounds on the bench press and pound out eight reps," Wickenheiser says, still amazed.

The summer of 2004 came to an end, but the hockey season of 2004–05 never returned. It was the NHL's lockout year. When hockey had last been played, in the spring, the Flames had made their run for the Cup. And when you win, or almost win, the party never quite ends until there is another winner. There's always someone around still wanting to celebrate.

Steve had been involved with the NHLPA since first coming into the league. A superstar has his own leverage; a players' union protects everyone else. The NHLPA is the players' team, and Steve was a good team player. The CBA between the league and the players would expire in September 2004. Until then, negotiations would follow their predictable path. Both sides would make respectful comments in public, and both would bad-mouth each other in private. Hatred and resolve would grow. There would be pretend talks during the summer, because the public needed to know that everyone was working tirelessly towards an agreement so the season could start on time. More importantly to those involved, both parties would need to demonstrate to their members—the NHLPA to the individual players; Bettman to the team owners—that they were doing everything humanly possible to extract the most from the other side and win the deal, which *can't* happen if agreement comes too early. In fact, it couldn't happen *if* the season started on time.

The beginning of the season would therefore need to come and go. There would need to be anxiety, then panic, and then with just enough time left to schedule just enough games to make a season of sufficient length to make the playoffs seem credible, a deal would be made. And the players would win because the players always win.

As it turned out, the season did not start on time. There was anxiety and panic—but then, when it came to the crunch, the negotiations went off script.

Both owners and players are competitors and both know how to win. But players have been teammates and team-players all of their lives, and know how to win together. Owners haven't been, and don't. In their own minds, players are team guys. In owners' own minds, they are self-made men. In this standoff with the league, NHLPA head Bob Goodenow only had to do what he had always done: say no to anything the league proposed, offer no alternative, and wait for the owners to crack and say yes.

Bettman had been losing to the players since he became NHL Commissioner in 1993. Player salaries had increased hugely and rapidly. To Bettman, this made no sense. It was up to each team's owner to make his decision, to sign a player or not, and to pay what he wanted to pay. Anti-trust law said so, and Bettman was always careful to remind the owners of that. Still, it was incomprehensible to him that owners could not stop themselves from getting into bidding wars over players. Sure, players could become free agents and go to other teams, but other players also become free agents and can sign with theirs. Why send salaries higher and higher? For owners, mindless competition for players was an always-losing proposition, Bettman believed, and only one team can win the Cup. Twenty-nine teams *will* lose. *Every year.* Bettman made this case to the owners passionately, incredulously, relentlessly. Any other conclusion was insanity. Any other action taken was clearly that of a fool. But then the first day of free agency would arrive each year on July 1 and all hell would break loose. Owners chased players, budgets were blown to smithereens. Bettman fumed.

It took him a decade and billions of dollars lost to the players to understand. Players are great competitors and need to win. Owners are great competitors and need to win, too. But a player, when he becomes a free agent, can go to five or six teams and still have a chance to win. An owner can win only in one place—where he is. So owners chased, players were chased, contracts went higher, and no amount of Bettman exhorting and fuming year after year was going to change that.

Left to their own devices, owners will break ranks and lose. Antitrust laws prevented them from working together, so Bettman got their support to take things out of their hands. He got them to agree to amend their own bylaws. It would require a simple majority— sixteen teams—to approve any proposed CBA deal that Bettman supported—but it would now take 75 per cent, or twenty-three teams, to approve any deal he opposed. He needed only eight teams on side to block any proposed CBA—a number he had easily in his back pocket. Now when Goodenow said no, Bettman could say no, too.

The CBA negotiation script held into October, and November, and December, but then when a deal was supposed to follow, it didn't. The season was cancelled. This time, the players cracked.

On July 13, 2005, a new owner-friendly CBA was signed. Two weeks later, Goodenow was fired; the NHLPA imploded. Bettman won, and he has never lost since.

The NHL's commissioner is an employee of the league. His authority rests with the Board of Governors. He needs their support; he himself can only influence. Bettman has always been adept at testing the winds and, if he finds them blowing the wrong way, shifting with them, or working patiently, calculatedly, to make them blow his way. He pretends that he must lead cautiously from behind. But with league revenues growing in multiples since his hiring, with fewer vulnerable franchises, and with most teams worth vastly more than in the decades before him—just as he did during the 2004–05 CBA negotiations, he has shown he can lead *forcefully* from behind when he wants to.

Steve worked out regularly during the early months of the lockout, but as with other players, without the usual urgency. If a day had to be missed or cut short, it was. There were NHLPA calls to participate in, meetings to attend. There were life details normally taken care of by the team that he had to attend to himself. Later in the season, he went to France to play for the Scorpions de Mulhouse. Mulhouse, once an industrial centre, the Manchester of France, is a picturesque and pleasant city of around 100,000 people, an hour's drive away from Zurich and Strasbourg. The Scorpions were a good team in a weak league. Steve played fifteen pleasant and picturesque games.

Elsewhere during the lockout, things were not standing still. Minor league players were playing minor league seasons, making themselves a year better and a year more ready to play in the NHL. Including Mike Commodore. Because of the similarity of their family names, he and Steve had been the two "Doors" who had come out of the press box during the 2004 playoffs to help the Flames to the Stanley Cup Final. Commodore had played twenty games in those playoffs, but had played few enough regular-season games not to be caught in the subsequent lockout, and he was now playing and developing in the AHL. Junior and college players were moving one year closer to graduation, to compete for jobs in the pros.

With the new CBA agreement signed, the Calgary Flames could plan for the 2005–06 season. They were no longer a team that had missed the playoffs seven straight years, in need of help everywhere. They were Stanley Cup finalists, and seemed only a few strategic moves away from a big, contending year. They also had Dion Phaneuf ready to join them. Phaneuf, drafted in 2003 from the Red Deer Rebels, had been named the WHL's best defenceman the following season, and would have played with the Flames in 2004–05 if not for the lockout. Instead, he stayed in Red Deer and won the top defenceman award for a second straight year. Phaneuf was strong and tough, a defensive and offensive force. He would score twenty goals with the Flames in his first NHL season, and finish third in the rookie-of-the-year voting, behind Alex Ovechkin and Sidney Crosby.

With Phaneuf's arrival, the Flames traded Commodore. Needing help on the power play, they signed free agent Roman Hamrlík; with Hamrlík, and young defenceman Mark Giordano, who had developed more rapidly than expected in the minors, they traded Denis Gauthier and Toni Lydman. That left Robyn Regehr, Rhett Warrener, Jordan Leopold, and Andrew Ference, along with Phaneuf, Hamrlík, Giordano, and Steve. After his play during the 2003–04 playoffs, Steve assumed he would contend for a top-four defence position, which would be a step up for him but, from the perspective of the now Cup-contending Flames, not necessarily for the team. Their top-four, even top-six, defencemen needed to be better. A top 1–2 pairing plays the big minutes, regular shifts, and on power plays and penalty kills—a total of about twenty-five minutes a game. A 3–4 pairing gets some power-play and penalty-killing time, and plays about twenty minutes in all. Just as a fifth starter in baseball gives you innings, 5–6 guys in hockey give you minutes, fifteen of them, maybe. They eat up time, giving the big guys a breather until they can come back on the ice again. A 1–2 guy's job is to win you games. A 5–6 (or seventh) guy's job is not to lose them. Steve improved the team by being a better seventh defenceman than they had imagined he would be two years earlier. But that's what he was. Then he had also been twenty-four, and seemed young. Now he was almost twenty-six.

For Steve, everything was confusing. In 2003–04, he had paid the dues he expected to pay and had always paid. Now it was payoff time, except that when the 2005–06 season began, he was in the press box even more than he had been before. He was surprised and disappointed. But he had felt that way before, and every other time he had picked himself up and worked harder. He was a team guy; he did what the team needed of him. If it wasn't to play at that moment, it was to get himself ready for the next one, to practise and train hard, to be a role model and inspiration, especially to the younger players—to help make them better, to make the team better.

But this time, Steve was practising and training with less purpose, and less pleasure. He couldn't stop himself from feeling this way, and he was upset and unhappy with himself that he couldn't. And if his on-ice life wasn't right, his life off the ice didn't feel right either. You have to earn the party after, and you can't play guilty. Maybe the drinking and drugs were holding him back, he began to think. He had always been a gamer. Whatever the situation, he would rise to the challenge, whether it was one that others had presented, or one that he had imposed on himself. If he were to stay up all night partying, he would still be in the gym at seven, drop 300 pounds on the bench press, and show everyone else, and himself, that he could do it. But this off-ice life didn't feel right if his on-ice life wasn't working. He was angry at his off-ice self. Everything began to feel shaky. Alcohol and drugs were putting his career and future in jeopardy.

In early December, after being a healthy scratch again, Steve went to see Darryl Sutter. He told the Flames coach what Sutter already knew: Steve was at a pivot point. He needed to play if he was to improve enough to stay in the league. Sutter agreed to trade him.

Calgary was on a ten-day, five-game road trip. They had lost in Nashville, won in Detroit and were on their way to Pittsburgh. Sutter met with Steve and told him that he had been traded to Florida. Rhett Warrener remembers the scene a short time later. The Flames players were on the bus in front of the hotel, about to leave for the airport. Steve was also in front of the hotel, waiting for a taxi to the airport to get to Miami. The bus left first. "And there was Monty," Warrener recalls, "just standing on the sidewalk with his bag." Warrener had played for the Panthers earlier in his career. In Florida, he knew, Steve would get a fresh start. He'd have a chance to find new interests and begin new habits. In Calgary, if a player did something wrong, everybody noticed and called the Flames office or contacted the media. In Florida, nobody saw anything because nobody was watching. Steve could hide in Florida. He could do anything. As the bus pulled out, "I looked back at Monty," Warrener says, "and I remember thinking, 'This could go either way.'"

Marty Gélinas, "the Eliminator," who had signed with the Panthers in the off-season, met Steve at the airport in Miami. He had been aware of Steve's drinking, but not the extent of it. When he heard of the trade, Gélinas and his wife sat down and talked. They agreed that Gélinas would suggest to Steve that he live with them, just until he was settled, of course. Gélinas knew what it was like to be in a new place: the Panthers were his sixth NHL team. "As a family, we didn't know how to do this," Gélinas recalls, "or how it would work. We had two young kids and my wife was pregnant. Sure we had question marks. But if someone needs your help to get back on track, to me it's a no-brainer."

Every morning, Steve left the Gélinas house before everybody else was up, but when he came back from practice, Gélinas recalls, "he would just be part of the family," as he had been with the Babcocks in Peterborough. "If something needed to be done around the house, he'd do it. He did the dishes. He played street hockey with the kids. He took them to hockey practice. We all sat down and had dinner together. We chatted at night, sometimes had deep conversations about a lot of different things. He became one of us." Steve stayed with them for the rest of the season. "We gave him some stability, and gave him a chance to see how a family man in a hockey family lived," Gélinas says. "How when you get home from practice maybe you've got to cut the grass, or take your kids to something. I think that year Monty realized there was more to this than just the game. There had to be some structure. There was a life. There was something he could shoot for. He made such an impact on our lives. It was really good for my family, and it was good for Monty. And the way he talked with my kids, one on one. It was awesome. And they were young. Not everybody can do that. They still talk about him."

Steve didn't dress for the Panthers' first game after the trade, nor for the second, a 6–3 loss at home to Ottawa. He played the

third game, on the road against Dallas. The predictable happened—
nineteen seconds after the Stars scored to take a 2–0 lead, Steve got
into a fight with Bill Guerin. Before the period ended, the Stars
scored again. The second period was different, however. The
Panthers scored twice, Steve assisting on the second goal, then tied
the game on a power-play goal by Joe Nieuwendyk, assisted by
Gary Roberts. Then, on a power play, with only sixteen seconds
remaining in the third period, the Stars scored and won the game,
4–3. As the game ended, Roberts—frustrated, angry—got a penalty
for boarding.

Nieuwendyk and Roberts had both grown up in Whitby,
Ontario—then a town of about 20,000, now a rapidly expanding
bedroom community fifty kilometres east of Toronto. At age six,
Nieuwendyk played tyke hockey for the Owls, Roberts for the
Wrens. Thirty-three years later, Nieuwendyk, having played more
than 1,200 NHL games, scored over 500 goals, and won three
Stanley Cups, and Roberts, having played more than 1,100 games,
and scored over 400 goals, and won one Stanley Cup, signed with
Florida. They were both thirty-nine years old.

They were signed to be mentors to a group of promising young
players—Jay Bouwmeester, Nathan Horton, Stephen Weiss,
Gregory Campbell—all of them twenty-two years of age or under.
Nieuwendyk and Roberts were to show them how winners—true
pros—practised, trained, played in big games, worked with teammates, treated fans, took care of themselves, and lived. "Both Gary
and I found it difficult," Nieuwendyk recalls. The two of them were
used to playing on teams that won most of the time, and always
contended. They had played in hockey cities, or on playoff-bound
teams that had made non-hockey cities, for a time at least, feel like
they were hockey cities. They had played where hockey mattered

and they mattered. They were also used to being better players, and weren't used to being old. At this stage in their careers, they needed energy, hope, and possibility; they needed more than their own professionalism and pride to drive them. It turned out Nieuwendyk and Roberts needed more from the younger players than what they were able to give.

"Monty was really a breath of fresh air," Nieuwendyk remembers. "He was fun and energetic." Older than the young guys, younger than Nieuwendyk and Roberts, Steve could relate to all of them and they could relate to him. He helped Nieuwendyk and Roberts connect to the rest of the team, so they could have the impact that the team needed them to have.

Steve was still a 5–6 defenceman getting 5–6 ice time—between fifteen and eighteen minutes a game, rarely on power plays or the penalty kill; he even played on the wing at times (it was this adaptability, as he told his Saint John coach Jim Playfair, that made him a million dollars). But now Steve was a regular; he arrived at the rink for every game knowing—not wondering if—he was playing. He liked his teammates. He liked the flip-flops and shorts informality of Florida. He liked living in the caring environment of the Gélinas family. He knew he needed to change his life, and now he was ready to do it.

In February 2006, the NHL schedule shut down for more than two weeks, for the Winter Olympics in Turin, Italy. During the much shorter All-Star breaks in other years, Steve might charter a plane with a few teammates and their wives or girlfriends, and go somewhere hot, or fly off by himself to some place he thought might be interesting. But this time it was different.

The Panthers played their final game before the Olympic break on February 11 in Buffalo, and wouldn't play again until the 28th.

Steve didn't tell his teammates; he didn't tell Keats, Jay, Nick, or Vally; he didn't tell Gisele Bourgeois or Marty Gélinas. His coach, Jacques Martin, and his general manager, Mike Keenan, didn't know. He didn't even tell Paul, Donna, Chris, or Lindsay until the day he left. He just got on a plane, flew to California, and checked himself into rehab. It was at Milestones Ranch in Malibu. He was there two weeks, as much time as he could risk in order for his rehab to go unnoticed. Then he flew back to Florida, resumed training when the team's practices began again, and continued with the season.

The Panthers' strength and conditioning coach, Andy O'Brien, didn't know about Steve's rehab stay, either. This was O'Brien's first NHL job, and he had known little about Steve before the trade from Calgary two months earlier. At first, Steve was just another guy who worked hard in the gym—and the guy who kept knocking over the hurdle. O'Brien had a runner's hurdle in a small area just outside his office. It was for the players to do stretches each morning, to open up their hips before going onto the ice to skate. The hurdle was made of metal, pieced together, and "if you hit it at all with your foot it would fall apart and make a pretty loud noise when it hit the floor," O'Brien recalls. "And every morning I knew when Monty was there because every morning I'd hear the hurdle hit the floor. Some days I literally counted the number of times he kicked it over. But every time he did, he'd put it back together and try it again. He had this clumsy, bit of a train wreck, element to him. It was very endearing, very sincere."

After the Olympic break, Steve and O'Brien grew closer. It began with Steve's workouts, then as he began to hang around, he and O'Brien began to talk. Their conversations grew long and intense. "I didn't really understand the significance of all this for him at the time," O'Brien says, "why Monty had this extra compulsion to drive himself as hard as he did." It was several more months before Steve told him about his rehab—and by then, Steve was starting to put some of his own pieces together in his mind. "He was beginning to make training a big element in the journey he was embarking on to change his life."

For Steve, it was hockey that had always made life work. But now, on this non-playoff team in this non-hockey market, surrounded by successful on-ice and off-ice people, he was beginning to realize that life made hockey work as well. He went to Alcoholics Anonymous every morning he was at home (that's where he was before the Gélinas family got up, something Gélinas came to know only later), and most days on the road. Going into rehab and to AA helped him give up one life; taking on the family habits of Gélinas, the playing habits of Nieuwendyk and Roberts, the wellness habits of O'Brien, and the life habits of all of them, would help give him another.

"For Monty, this wasn't just about changing one or two behaviours. It was about embracing a new way of living," O'Brien says. "He was giving up an old obsession and gaining a new one. The way he exercised, and respected his sleep. The way he respected his nutrition. He could see an element of purity in the way he was living. He started to realize that what he put into his body and what he did the night before might overcome anything he did in his next day's workout. So he didn't want to put anything toxic in his body at all. This wasn't about dealing with addiction. For him it was about adopting healthy living."

And as always, once Steve was in, he was all in. He wanted to know everything. "If he heard about something, he'd ask me my thoughts, we'd debate it, and he'd want to try it. And he also wanted to explore things on his own, have his own experiences, know for himself." He talked to everybody and tried to learn from everybody. He wanted to be better, and he wanted to discover and uncover and maximize everything that was in him as a hockey player and as a person. For Steve, this wasn't overcoming; this was pursuit. This wasn't grimness or desperation; this was joy. Possibility. He couldn't wait to get to practice, to get to the gym, to do every rep he was supposed to do, and ten more. He couldn't wait to get to the health-food store and the grocery store. He couldn't wait to meet the next person; he couldn't wait until he could be off by himself to read or think. He couldn't wait until the next day. Some of his Peterborough friends were worried about him, about this "fanatical pursuit of health," as O'Brien puts it. It felt manic. It had a hint of instability about it. O'Brien disagrees: "I can say with certainty that at the times I knew him, from Florida to late in his life, he was a very, very healthy person."

And in finding this new life, Steve appeared little haunted by his old one. He didn't seem to have days where, so in need of a drink he hid himself away to avoid the temptation. He went with his old friends to bars and restaurants where they drank, and he didn't. He didn't go as often as he did before, nor did he spend as much time in Peterborough during the summers. Still, his friends were his friends, and his old friends were still as funny and like-able even when he was sober, and he was still as funny and likeable to them. He liked his new self. He liked it at least as much as he'd liked his old one, and his friends did too. Steve's new life "seemed all very comfortable and natural to him," says O'Brien.

Steve was also finding a new life in reading. The book that had the greatest impact on him was *The Four Agreements* by Don Miguel Ruiz. Steve read it and reread it; he highlighted sections of it; he kept it with him on the road. First published in 1997, *The Four Agreements* was on the *New York Times* bestseller list for more than seven years. In it, Ruiz relates his own story. Born in Mexico, he almost died in a car accident while he was attending medical school. Yet, as damaged as his body was, his mind remained fully alive. At that moment, Ruiz writes, he understood what for him became a simple truth: "I am not my body." Later, as a doctor treating patients with all manner of illnesses and injuries, this truth took on greater meaning for him. To help them heal their bodies, he realized, he had to help them heal their minds.

The "Four Agreements" he details in this book, Ruiz says (though others have disputed this), come from the spiritual traditions of the Toltecs, a people who inhabited parts of south-central Mexico more than 3,000 years ago. Ruiz believes that every child is born as a blank page, free, and with every human possibility in front of them. But the world that they are born into—of parents, families, schools, cultures, societies—has already determined countless understandings about what they are and how things work, which every child learns, both by reward and by punishment. By this process of "the domestication of humans," as Ruiz calls it, the child comes to live everyone else's life, not his or her own. The book is about how we can live a life that is centred in ourselves—that is our own. It is, as it describes itself, "a practical guide to personal freedom," detailing four "agreements" to live by.

> *The First Agreement: Be impeccable with your word.* The word *impeccability*, Ruiz explains, comes from the Latin prefix *im-*, meaning "not," and *peccare*, meaning "to sin." Even if individuals and institutions in the world around us are not speaking and living out their own truths, we must know our truth—in order to express it and to live it. We must speak about others and about ourselves "without sin." This First Agreement, Ruiz

says, "is the most important one and also the most
difficult one to honour."

The Second Agreement: Don't take anything personally. When
others say something about you, or do something to
you, it's not about you. It's about them living their
lives and filling some need they have, which can be
hurtful to you if you allow it to be.

The Third Agreement: Don't make assumptions. We assume
everyone sees the world the same way as we do.
They don't.

The Fourth Agreement: Always do your best. This is the action
you take to realize the first three agreements.

And if you do live up to all four of these agreements, Ruiz says,
"You are going to control your life one hundred per cent."

The Four Agreements is a harsh message in a gentle story. Nobody
and nothing in the world is what they should be. Everybody's truth
is false. The path to your own life is not about forgiving others;
that makes them too important to your life. The path is about
becoming, each of us individually, in control of our own life. It was
Gisele Bourgeois who had given Steve the book. She wanted him
to know that in a life his parents had first lived for him, then his
parents and coaches, then his coaches and managers and the media
and fans, that it was *his* life. He had to shut out the thousands of
voices around him and hear only his own. He had to know that
even if he was a team player, it wasn't because others—the players,
coaches, and fans—said so; it was because *he* said so. Because he
said *impeccably* that this is what he was.

The Four Agreements meant so much to Steve, O'Brien believes,
"because Monty was a guy who had a lot of hope. He had faith in

what was coming around the corner. He understood that no matter how hard things are, you can always get through them. That drove him to find solutions, to seek out better opportunity no matter what situation he was in in his life. I know there was never a moment he didn't believe in himself. Or that he didn't believe he was capable of getting to the next level. That's so important for a guy who's in the American League, then called up, then sent down, then a healthy scratch; moving back and forth from defence to forward. He always believed he was capable of getting to that next level. In anything."

In the 2005–06 season the Panthers earned ten more points than they had the season before the lockout, but missed the playoffs for the fifth straight time.

Steve went back to Calgary that summer. Hayley Wickenheiser recalls a conversation they had one day after a workout in the gym. "Monty asked me to sit in his car and he told me that he had been to rehab. He just wanted me to know that, he said, and he wanted me to know that he was an alcoholic and that he was going to change his life. I remember being taken aback, but he was attempting to do his cleanse."

Steve did the same with Keats. "I was working in Hoboken, New Jersey, at the time," Keats says. "I remember him calling and telling me that he was an alcoholic." He sounded serious, but Keats knew what would follow, what had always followed with Steve: an even bigger than usual Steve-laugh, because he had sucked Keats in again, and that Keats would never hear the end of it. Keats didn't realize that something might be different this time until Steve started talking about "making amends." They were buddies. Buddies don't need to make amends to buddies. Keats had been as drunk and stupid with Steve as Steve had ever been with him. "I remember saying to him, 'You don't have to do this [make amends] to me.' Then we laughed a bit. I actually said I was proud of him for doing this."

The Panthers got off to an indifferent start to the 2006–07 season, then had a bad November, winning only three out of thirteen games. They had a new goalie, Ed Belfour, who at forty-one was giving the team the same strong goaltending that Roberto Luongo had provided for the previous five seasons. Luongo had been traded to Vancouver in the off-season in a multi-player deal that included Todd Bertuzzi, who the Panthers believed would give the team the scoring and attitude they needed. Instead, after recording a goal and three assists in his first game, Bertuzzi injured his back a few games later, had surgery, and never played for the Panthers again. Joe Nieuwendyk, also with recurring back problems, played only fifteen games, went on injured reserve, and retired.

The Panthers improved during the second half of the season, but they had too much ground to make up. Their veterans were too old or too injured, or both; Gary Roberts went to the Penguins at the trade deadline. The young guys were too young or not good enough, or both. The Panthers needed more than what their players could give them, Steve included.

Steve was now twenty-seven years old. He had been a solid player and a good teammate all his life. At times, and to some coaches' or managers' eyes, he had been more or less than that—or would become more or less. But really, truly, after all these years—through youth and adolescence, substance abuse and sobriety; after all his training and learning, hope and hard work—he was what he was going to be. No less, but no more.

Nieuwendyk, for almost twenty years, had played with good teams against good opponents in the NHL and at the Olympics. He knew Steve as an opponent, a teammate, and from watching him play. "Monty was always a gamer," he says. "He liked to be involved. He wanted to help the team. He was physical and gritty. He played hard. He wasn't a fun guy to play against. And he wasn't very big, but he would do anything for a teammate. Stick up for him. Drop the gloves. He was a good skater, and could carry the puck. He was a good first-pass guy. It's just that when the walls were closing in on him he tried to make a harder play than he needed to." If he was playing against him, Nieuwendyk says, he'd make the walls close in. Get on him quickly, cut off the passing lanes, turn Steve's hard-trying instinct against him, get him to do something he couldn't do. "He would have been better served at times to just play the simple game.

"I think he wanted desperately to be an everyday player. He wanted to make an impact. To be one of the core guys. Monty really searched for that kind of respect and stability. I think he chased it throughout his career. He was a great 5–6 pairing guy, but he was never going to be anything other than that." Steve was the type of player that moves around, and goes from team to team. "He didn't have to worry about family," Nieuwendyk says. If he was traded, or became a free agent, "He was a guy that could say, 'Okay, gotta close up the apartment. I'm off to the next stop.'"

On the ice or off, Steve could fit in everywhere, and not make a home anywhere.

Late in the season, Steve celebrated a milestone. "Monty invited me to his one-year anniversary of being sober," Gélinas says, "where AA gave him that medal. It was pretty special."

CHAPTER THIRTEEN

Steve's first season with Florida had been Keith Primeau's last season in the NHL.

Primeau was big—six-foot-five, 220 pounds—and he moved with the power and fluidity of a big Lab, always on the puck or always on the player with it. Detroit's first-round pick, the third selection overall in the 1990 draft, he played for Canada in the 1996 World Cup and in the 1998 Olympics, and was captain of the Carolina Hurricanes and the Philadelphia Flyers.

In 2004, the Tampa Bay Lightning had beaten Primeau's Flyers in the seventh game of the Stanley Cup semifinals before going on to win the final against the Flames and Steve. In a radio interview, Phil Esposito described Primeau in that series as "the most dominating player I ever saw. More than Orr, Howe, Gretzky, or anyone." During his career, Primeau played the way a Canadian player of the time was supposed to play—tough, competitive, unrelenting, with a crasher's spirit and a goal scorer's touch. On October 25, 2005, his career ended after he received the last in a series of concussions.

"I had four documented concussions," Primeau says, careful to distinguish the documented ones from all the other times he had his bell rung as a kid or in junior, or even from the time in the 1996 World Cup when he ran into his own teammate, Eric Lindros. "[Lindros] circled one way in the neutral zone, I circled the other," Primeau explains, "and we collided." Primeau is quite certain now that this injury in the World Cup left him vulnerable to the concussion he received in the season that followed. That, too, came in the neutral zone, and wasn't a severe hit. But as Primeau recalls, "It was enough to put me in the hospital overnight," and for the team to shut him down for a week. He searches for the right words to describe how he felt in those moments. Overwhelmingly, he says, "I just didn't feel well. I didn't feel myself. I didn't feel normal. I wasn't processing as quickly. I wasn't thinking as alertly. I was more tired than usual. Physically, my eyes felt heavy. Everything seemed slowed down. I had a lethargic perspective on everything. I wasn't stimulated as easily. I became more irritable. I had less patience. Instead of getting up in the morning feeling good about the day, about what I was going to get accomplished, I was just inside my body trying to function."

Everything in his life "felt very, very delayed," Primeau says, like when a DJ slows down a record on his turntable and the sound *commmes ouuuttt liiikke thiiss*. All the while, he says, "I was trying as best I could to do other things, because that allowed me to be outside of my own mind, so I wasn't thinking about how bad I felt."

"I can remember times when I played concussed," Primeau continues. "I'd be hit, and come off after a shift and be thinking, 'When can I go back out there?' Because when I was focused on the game, it didn't allow me to be where I didn't want to be." When he was playing, he didn't have the chance to think about how bad he felt. When the game was over, Primeau would find he didn't feel better, but he didn't feel worse. Not playing was harder. "On the exterior, you don't look as if anything is wrong with you," he says. "That was the most difficult part, walking into a locker room with a bunch of guys, and you look perfectly fine. You're speaking perfectly fine. But you're not fine." And they were playing, and he was not.

Primeau's second documented concussion occurred in May 2000. "I was taken off the ice on a stretcher in Pittsburgh," he recalls. "That was probably the most severe concussion I suffered, and it was probably the beginning for me of what I call 'the demise,' because I didn't manage it the right way. I was back playing two nights later when I was clearly suffering post-concussion symptoms."

The concussion happened in the second round of the playoffs against the Penguins, their in-state rivals. "I won the faceoff in the neutral zone," Primeau recalls, "and our defenceman threw a how-itzer [pass] at me." It was behind him. "As I turned to look for the puck, their defenceman stepped up and blindsided me." It was Bob Boughner; he caught Primeau with a shoulder and elbow to the head. "In today's game, it would probably be a twenty-game sus-pension," Primeau thinks. "You could say that I had my head down, but my head wasn't down." Primeau didn't have the puck.

"My first instinct was to get up," he recalls, "but they wouldn't let me. They didn't know if I had any cervical [neck] damage. That's why they took me off on a stretcher. But I felt fine. I didn't have a headache. Maybe I felt slowed down, but I had none of what you consider to be concussion symptoms—you know, headache, head pressure, nausea, blurred vision, any of that stuff." They took him to the Flyers' dressing room, and then to the hospital, where he was kept overnight for observation. The team flew back to Philadelphia. When Primeau woke up the next day he still felt fine. The Flyers' owner, Ed Snider, sent his plane to Pittsburgh to fly Primeau back to Philadelphia. There, he was given a baseline test. "We had done baselines so many times before that I knew what the tests were going to be. So I went in there very focused and prepared, and I passed it." Afterwards, the neurologist that the Flyers had on their staff told him that it was her responsibility to give him the test, that she had done that, and the test said he was fine. "Then she looked at me and said, 'Keith, just be very careful. Make sure you know what you're doing.' I kind of giggled," Primeau recalls, "and I said, 'I know. I'm fine. I know what I'm doing.' And then I walked out of her office and I had the worst headache I've ever had in my life. It was this searing pain, because I'd focused and concentrated so hard just to pass my baseline, to get permission to play." The next game was two days away.

Primeau was a player. He knew that he could have a headache one day, and the next day, and even the day after, and right up until game time—and none of that mattered. "The mentality is: 'This headache will pass,'" he says. It had always done so in the past, so surely it would in the future. He was certain he "wasn't doing irreparable damage to my brain by trying to persevere. It wasn't that big of a deal at that point."

Primeau didn't go on the ice that day, but he did go on the next, the day before the game. "I wasn't a hundred per cent," he recalls, "but I felt well enough to say, 'I'm good to go.'" The worst of the headache was gone, and it was still thirty hours until the puck dropped. Thirty hours later, he felt no better. But anybody can play healthy. Primeau was a competitor; this was another test. Besides, the team needed him. "I thought that by playing I was motivating my teammates," he explains. The Flyers won the game 4–3 and the series in six games.

Next, the New Jersey Devils. Primeau was feeling no better, but no worse. In the middle of the second period of the second game, the Flyers behind 2–1, Primeau got into a fight with the Devils' Randy McKay—"not a guy I thought could seriously hurt me," he says, "but a tough kid and a smart fighter." Primeau wanted to show his teammates that he was so committed and so "okay" that he was willing to fight. In Game 7, the Flyers' Eric Lindros rushed the puck up the ice as if he were still playing peewee—too big, too strong, too good to be hit; but lurking in the shadows was the Devils' defenceman, Scott Stevens. When Lindros was completely defenceless, Stevens obliterated him with a shoulder to the head, ending Lindros's career as a superstar. The Devils won the series.

It was May 26. Primeau had three and a half months of warmth and sunshine ahead of him before training camp began, a time for his bumps and bruises to heal. That's how it had been in other years, for his knee and his back; why not for his head? With no big surgeries to rehab, this would be a normal summer for Primeau.

And it was. When the next season began, Primeau played the way he had before his injuries. His reactions, his hands were just as quick; he felt entirely himself. "I was fine," he said, "until I suffered my third documented concussion." It came almost four years later, in February 2004.

Again, Primeau was in the neutral zone, and like a wide receiver cutting across the middle of the field, his eyes were on the puck; the defenders' eyes were on him. "I got hit on the left side of my head with a forearm and elbow," he recalls. It was from the Rangers' Bobby Holik. "I didn't think it was that hard," he said, but this time, he immediately didn't feel right. "They took me to the locker room and told me I probably had a concussion. And I said, 'No, I'm fine. I feel fine.' But there was something about my eyes,

I guess. Even my wife when I got home said the same thing." This was a Thursday night at Madison Square Garden; the Flyers' next game was Saturday afternoon at home.

The next day, Friday, Primeau was driving to practice and was only a mile away from his house. "All of a sudden my stomach was nauseous," he said. "I pulled into a parking lot and called the trainer. He told me to go home and get into bed and come back to the rink the next day. It was delayed onset." The playoffs were due to begin in eight weeks.

For the first time, the problem for Primeau was more than just headaches and not feeling right. "It was motion sickness," he recalls, "bright lights, and I started to struggle with my speech." Not so much that others noticed, but he had to work harder just to get his words out, and felt no real improvement day to day. But week to week, slowly, would be fast enough, he knew. He needed to get back for the playoffs. He set in his mind the last weekend of the regular season for his return. In those final games, he would be able to get in a little conditioning, get back some of his feel for the game, and be ready when he needed to be. In the last week or two, "I started to feel well enough to convince myself and others that I was okay to play. My eyes, my vision, were still off. But this wasn't about feeling perfect, this was about those three-hour windows of a game, when everything goes away but the game; to have that outlet, just to feel well enough to play. That's what I was fighting for."

Primeau hit his target, and in eighteen playoff games that year, scored nine goals. This was the year the Flyers went to the semifinals and lost to Tampa Bay; when, because of his play in this series, Esposito called him the most dominating player he'd ever seen.

"I had a good playoff run," Primeau says. "And I think it was partly due to how hard I had to concentrate and focus." By the time the playoffs were done, "I thought I was healed."

Two nights after their series loss to the Lightning, the players went to the Borgata casino in Atlantic City. "It was an end-of-the-year party, it was all the guys out drinking and blowing off steam. I had a fair bit to drink, but nothing more than I'd had before." The next morning, he recalls, "I couldn't get out of bed. I couldn't get my head off the pillow. It felt like the whole weight of my body was in my head."

For five days, nothing changed. Then he started to feel better. His symptoms receded. "I got back into my routine, got the kids off to school. We were into the summer, and I was going to take a few weeks off anyway. And we were going into the lockout year. Everything would be fine when it needed to be fine. There was plenty of time to rest and recover." Instead, his symptoms were slow to clear, and if training camp had begun in September, he wouldn't have been ready to play. But as it turned out with the lockout, Primeau had an extra year to heal; an extra year to seek out every treatment imaginable.

One year later, in September 2005, he was at training camp. He felt "good and recovered," Primeau recalls. "[Then] we were seven games into the season and everything is going fine. We were in Montreal, I'm out killing a penalty, our goalie kicks the puck out, I go to clear it away and [Alexander] Perezhogin comes flying in for the rebound and catches me in the side of the head. I go down and stay down, just waiting to see how I feel. And I felt fine. There were only about seven minutes left in the game, so they didn't play me anymore. After the game I got on the bus and called my wife and my parents, and I was excited. I said, 'I'm fine. I took a hit. I feel good. I think we're good to go.'"

The team returned to Philadelphia. Primeau practised the next day, played against Florida, then flew to Carolina and was in the lineup against the Hurricanes, but he didn't feel right. "I was sluggish, lethargic, my legs were heavy, I just didn't have any jump. But I thought, 'It's early in the year.' I still wasn't associating how I felt with the hit in Montreal." The team flew to Ottawa after the game and practised the next day. The morning after that, on game day, "I get up and go to the rink. I still don't feel great during our skate, and go back to the hotel but I can't sleep. Then I go and see the trainer, and they keep me out of the game." Primeau had felt tired and listless in the first days after being hit in Montreal; but then, as he put it, "things deteriorated: the sensitivity to light, double vision, headaches, head pressure, exercise-induced light-headedness, clearly the worst symptoms I'd had to this point."

When the team got back to Philadelphia, Primeau went home and took a couple of days off and waited for the symptoms to subside; but they didn't. He started into his recovery routine. "I would go to the rink every day, ride the bike, get light-headed, go home and sit in front of my computer in a dark room, and not feel great. Then I would go to the rink the next day, and feel the same." Weeks passed. "I started seeing a bunch of different doctors, and tried different therapies: massage, acupuncture, reiki, rolfing. I was doing vestibular therapy. I was going to downtown Philly for visual therapy at least twice a week, and would come out of there completely exhausted because of how hard I stressed my eyes. I was willing to give anything a shot. That was my cycle for the better part of twelve months. I had some days that were slightly better than others, but no good days. None were symptom-free."

He skated with the team a few times near the end of the 2005–06 season, but then people began to think he might be back for the playoffs, and, rather than becoming a distraction for the team, he stopped. He went home for the summer. "The team figured we had done everything we could here, and so why not rest and try and get better." Instead, Primeau recalls, "I had a terrible summer. The double vision had gone away, and most of the exercise-induced light-headedness, but I wasn't getting better." In late August, he returned to Philadelphia and started skating with his teammates. One day when their workout session ended, Primeau went into the trainer's room and said to Jim McCrossin, the team's head athletic trainer, "I'm right there. I'm feeling good. In another couple of weeks when camp opens, I'll be fine." Primeau said this to McCrossin knowing he was not. At this point, McCrossin stopped him. "Keith," he said, "we appreciate your effort and the work you've put into getting back. But in good conscience, I'll never be able to give you permission to play again."

"At that point," Primeau recalls, "my career was over." Was he angry? Lost? Devastated by what McCrossin had said? "I felt this sense of relief, because I would have just kept pushing." But the hard part was only beginning. "Just because you retire," Primeau says, "doesn't mean you're healed. My symptoms lasted for another seven years."

Nothing really changed, month after month, year after year. "I was always just tired, and not really able to do anything for any extended period of time. The headaches and the head pressure; I couldn't exercise." He had skated, worked out, done something physical almost every day, all his life. Now—nothing, not even with his kids. "Even getting down on the ground and playing mini-hockey with them, I couldn't do it. The movement of my head would set me off."

He remembered how he had been when he was healthy, and that haunted him, and it drove him. "I knew what normalcy was," Primeau says, "and I had too much to live for not to fight to get back to it. So I dealt with the headaches every day." He had all these years of life ahead of him, and he wasn't going to accept the way he felt as his new normal. He would do everything he could, see every doctor, take every treatment, to get himself back to his old normal—but in the meantime, he had his life to live. He had a wife and kids. It had to seem to them, headaches and soul-weariness notwithstanding, that he was never anything other than the old normal.

"And this is where it gets very delicate," Primeau says. He mentions Dave Duerson, Andre Waters, and Junior Seau, among others—football players who, after suffering head injuries and their dehumanizing effects, had killed themselves. "We all have a breaking point," he says. "I wasn't at it, and I don't know what mine is. But there had to be a point where I'd stop caring about getting back to normalcy and would spiral out of control. Because there were days where I sat in my office with the lights off, breaking out in tears, because I'm six, seven years into this and I don't feel well.

"I lived on hope. I hoped to return to my normalcy, and I was going to fight every day to get there."

Primeau had to adjust; think about things differently. He had to focus on what he was capable of doing, not what he wanted to do. "It meant that if I got tired and had to take an afternoon nap, I took an afternoon nap. If I wasn't able to work out, I didn't work out. I laid down a lot," he says, "because that's when I got the most relief." It might mean a couple of hours in the afternoon, or half an hour at other times of the day—just little compromises that gave him some respite. "Basically, I let my brain dictate what I was going to do,"

Primeau says. "If I needed to take a nap, it was because my brain was telling me I was tired."

About two years ago, his symptoms receded. They had been diminishing gradually, so much so that he hadn't really noticed, until one day he thought, "Wow, I feel great. I feel incredible." Running still bothered him, but "I was working out and lifting again," he says. "I was training with my daughter. I lost a bunch of weight. I was still suffering from some fatigue, but it wasn't as excessive as it had been."

Then, a few months after that, he was at a rink for one of his kids' practices. He was in the dressing room and walking to the ice, but at six-foot-five and wearing skates, he forgot to duck, and hit his head on the top of the doorframe. His symptoms returned.

They lasted about a year: headaches, dizziness, a shortening of his attention span. Then, beginning around September 2015, he could feel them go away again. He still isn't able to run any distance. The moments of creeping tiredness that he feels may always be with him. "I'd love to be at a hundred per cent again," he says. "But I'm okay with where I am."

His old symptoms may be only a random bump away, yet Primeau knows now that those symptoms can also go away again— because they did. His old normal is still in him. "I don't think I was close to my breaking point," he says, "but I certainly can appreciate when somebody is. I think it is important when we lose hope that we're ever going to get better, that we just keep fighting."

CHAPTER FOURTEEN

Gary Bettman emerged after the 2004–05 lockout in complete command. He got his salary cap and, in the process, got the NHLPA to destroy itself.

The NHL now had seven years of labour peace ahead, seven years of wage certainty, seven years to construct a solid base on which shaky franchises could stabilize—and strong ones, as well as the league itself, take off. Only one team moved in the decade that followed: the Atlanta Thrashers became the Winnipeg Jets. When the Jets had left for Arizona in 1996, there were few things more certain than that Winnipeg would never have an NHL team again. The city's population was less than one million. Its corporate base was too small; its traditional industries were in agriculture. Once the "Gateway to the West," Winnipeg was losing its importance in a country that was losing its importance in the NHL. Of more significance to the league, the city offered little upside to an NHL that was selling upside. For U.S. franchises needing to convey "bigness" to their arena- and TV-publics, no city name on a marquee said "small" more than Winnipeg. But in 2011, seemingly impossibly, the Jets came back. It was as if the Dodgers had returned to Brooklyn. A city that had once seemed bigger than it was because it had an NHL team—then came to seem smaller because it didn't—began to feel bigger again. And if this could happen to Winnipeg, why not to Quebec and the Nordiques?

The mood of Canadian fans, in general, was beginning to pick up. Canadian teams were doing better. The value of the Canadian dollar rose; teams could afford to keep their players and compete against U.S. teams to sign others. There was a salary cap, which went some way toward levelling the playing field. Teams in bigger U.S. markets couldn't spend that much more than Canadian teams— and, as hockey markets, not many of them were much bigger any- way. No Canadian team had won a Stanley Cup since Montreal in 1993, at the beginning of the low-Canadian-dollar years, but the country's three smallest-market teams—Calgary, Edmonton, and Ottawa—were in three straight Cup Finals from 2004 to 2007. Canada won the gold medal in the 2002 Olympics, lost in 2006, and hasn't lost since. And for the first time in many years, Canada had the best player in the world, Sidney Crosby. Strong Canadian teams were making the NHL stronger and feel better about itself.

More than a century after hockey had moved indoors, it moved outdoors again in 2003 for a regular-season game at Commonwealth Stadium in Edmonton, between the "team of the past quarter- century," the Oilers, and the "team of all-time," the Canadiens. It was a frigid, prairie-clear winter afternoon. Canadiens goalie José Théodore wore a toque; you could see the players' breath. It was magical. Players who normally complain at the slightest imperfec- tion of the ice surface looked like little kids on a very imperfect back pond having the time of their lives. Their joy came through the screen. Canadian fans *got it*; so did sunbelt American fans. It was like watching Abner Doubleday and his friends on a pasture in Cooperstown; like watching a long-imagined legend made real. The Edmonton game led to many more outdoor games in the following years, each one greatly anticipated by the fans in the stadium; each hugely enjoyed by the fans at home.

These outdoor games earned the league increased TV attention. There were also now more all-sports networks with time to fill that needed programming. Using new technologies, any show could be watched at a viewer's convenience; but sports were live. Ratings went higher; sports programming became more valuable. In the late 1990s, the NHL signed big new network contracts in the U.S. and in Canada.

Hockey was also gaining traction in the U.S. In 1965–66, 1.6 per cent of NHL players were from the U.S.; 97.7 per cent from Canada. Twenty-five years later, in 1990–91, 16.7 per cent were American, a tenfold increase; 73.6 per cent Canadian. Today, less than half of all the NHL's players are from Canada (46.4 per cent); just over one-quarter is from the U.S. (26.5 per cent). American fans always had their teams; now they had their own players to cheer for. Hockey was no longer exclusively a Canadian game.

When the players returned for the 2005–06 season after the lockout, they had new rules to adapt to. The red goal lines that extend across the ice were now two feet closer to the end boards, and the blue lines two feet closer to the centre line, giving attacking players four additional feet in the offensive zone to find open ice; to elude defensive players trying to shut them down. As a consequence, the distance between the blue lines was reduced by four feet, which would have constrained the game in the neutral zone where it normally speeds up, except that passes were now allowed from the end boards in one zone, over the centre red line to the far blue line, a distance of one hundred and twenty-five feet—twenty-five feet further than before. This meant more open ice, better opportunities for players to pass, and more space for them to pick up speed.

The "tag-up" rule was also reinstituted, allowing attacking players to re-enter the offensive zone onside; and goalies were penalized for delaying the game if they "froze" the puck unnecessarily—both rule changes intended to keep the game moving. For defenders and coaches looking to slow the game down, to offer some safety to a defenceman retreating into his zone for the puck with a checker speeding after him and hunting him down, the league's referees were given the following directive: There would be "zero tolerance on Interference, Hooking, and Holding/Obstruction."

One other change had a remarkable, unintended effect. The NHL hates tie games. Baseball and basketball didn't have ties; ties are infrequent in football. But scoring is so difficult in hockey that if a game is allowed to be extended into overtime without a time limit, as in the playoffs, it might go on forever, which is a big problem when a team plays so many games a week and TV networks have schedules that viewers depend on. In 1999, the NHL had introduced a new rule: the five-minute overtime for regular-season games was now played four players against four; with more open ice, a goal was more likely to be scored. Yet many games still ended in a tie. So, six years later, the league decided there would be a shootout for games that were still tied after overtime. Critics were not happy. Four-on-four overtime, at least, was hockey-*like*; the shootout, borrowed from soccer, was a sideshow novelty whose only merit was that it guaranteed a winner. It took several months for teams to realize that, in a league of extreme parity, this "sideshow," which also happened to earn the winning team an extra point, might make the difference between making the playoffs, a player earning more money, and a coach keeping his job—or not.

This is where the unintended effect comes into play. Since that very first game at McGill, hockey players stayed after practice to work on their own special, outrageous moves that they had spent hours dreaming up; and coaches spent hours telling them to stop wasting their time on something they would never use in a game. But now there was the shootout. Goalies quickly came to learn even a shooter's best moves, which meant a shooter needed to surprise the goalie with something he had never seen before. All that post-practice time and the hours of dreaming up tricky moves suddenly didn't seem so dumb, even to coaches. And because of all-sports networks, the best of the best of these moves were showcased every night to fans—and the next day, to players too busy to watch them the night before—triggering even more outrageous moves. This was also at a time when shooters were finding it harder to score, because goalies were filling more of the net with bigger equipment and a new, butterfly style of goaltending suited to their now-oversized dimensions. With practice, the players got better at the shootout, and began using their moves not only in the controlled environment of the shootout, but in the frenzy of a game. And again, players would be glued to the next day's highlight clips.

What had begun as a sideshow changed the game.

The players also realized that if goals were more likely to come from creative new moves than from long-range shots that goalies just smothered with their Brobdingnagian-sized padding, they needed lighter sticks to give them the precision required to take these shots. And lighter sticks, coincidentally, also allowed them to pass more accurately—which meant the game moved faster, and shifts got shorter. There was even less coasting, less circling; it was just moving—from the instant a player's skates hit the ice to the instant they left it. One shift, the next, the next, the game a sixty-minute relay race, the baton passing every forty seconds.

Since rosters had increased in size in the 1960s, teams had run four lines. A scoring line, a scoring/checking line, a checking line, and a line of "specialists"—energy guys; bangers and fighters. Coaches tried to match lines against their opponents, to put their checkers against the other team's scorers, to keep their scorers away from the other team's checkers. Now the shorter shifts were too fast to control; even a ten-second mismatch meant having the wrong players against the wrong players for one-quarter of a shift. And what if the other team changed the order and threw out their scorers while your specialists were still on the ice? The puck would go quickly into your zone; you wouldn't be able to get it out, or change. The result was bangers chasing, and fighters with nobody to fight. Coaches love fighters because they put their all so visibly into the game. Teammates love them because they fight for the team. But if having fighters on the ice hurts the team, coaches and teammates don't love them that much. A game that is a sixty-minute sprint needs all four lines playing; and the fighters can't keep up. Somehow, in this game played in an enclosed space, that moves so fast that fights are inevitable, fighters have become obsolete. That decades-long explanation of hockey's exceptionalism regarding fighting turned out to be a crock. And now, as specialists have been replaced by fourth-line players who can skate and play, and checkers can't be matched up against scorers, the game has speeded up even more.

There are signs of this accelerating speed. The charging penalty has largely disappeared. It was a feature of a coasting game. It recognized the vulnerability of a puck carrier when an opponent took a few quick strides towards him and struck him with the full force of his body. But a full-sprint game is not played in a few quick strides, then a few more; it is played in ten, then ten more. It is all quick strides, so everything is charging, or nothing is charging—and the force and the frequency of collisions increase, with little to hold them back.

It is also a game too fast *not* to finish your checks, it seems. A player without the puck cannot be hit—that would be interference—because he doesn't have the puck, and if he could be hit then any player on the ice could be hit at any time, with or without the puck, which wouldn't make sense. But if a player doesn't have the puck, but he did have it not *too* long before, he *can* be hit, even though now he is really no different from every other player on the ice without the puck—except somehow he is, because that's what the rules allow. So he can be hit, even though he succeeded in passing the puck before the checker arrived—and the checker can hit him even though he failed to stop him before he made the pass. Furthermore, the checker, who failed, isn't penalized, and in fact is rewarded, because he has taken his man out of the play, and the passer, who succeeded, *is* penalized, because he cannot continue up the ice with the play, and can get blasted by the checker and because he, the passer, who in making the pass now has no time to protect himself, "should have had his head up" and "should have seen the checker coming." Because in hockey if you have your head down, you deserve anything you get—and anything you get, in this case, is called "finishing your check."

Many players have had their careers shortened by injury. Some by eye injuries—Doug Barkley, Barry Ashbee, Pierre Mondou, Bryan Berard; some from knees—Bobby Orr, Pavel Bure, Cam Neely; some from back injuries—Mario Lemieux, Mike Bossy. Many more had their careers diminished by injuries. For others still, it was a combination of age and injury. In almost every case, the reason was specific—a hit, a stick, a puck—something careless, unlucky, weird, even dirty, but each with its own individual explanation, nothing that suggested a larger problem, nothing systematic to the game. It was instead something unfortunate, sad, that somehow just happens not because it's hockey, but because it's sports. It's life. It was the same for head injuries, even the destruction of Ace Bailey by Eddie Shore, or of Ted Green by Wayne Maki. These incidents should never have happened, but they were one-offs—an individual player "losing it" in a dangerous way.

And even if there were signs, the signs seemed only enough to make you notice, to wonder a little, not to make it into a major priority. There were so many other things to think about—the value of the Canadian dollar, the competitiveness of Canadian teams, the empty seats in many U.S. sunbelt cities. TV contracts. In 1997, the NHL and NHLPA created a Concussion Program made up of medical experts that they each appointed. Three years later, they formed an Injury Analysis Panel of team doctors, trainers, GMs, referees, players, and league and NHLPA officials. But these were contentious times. The CBA was set to expire, a lockout was looming, whatever the league thought important about anything, the players wondered why, and didn't. And vice versa. Panel members might find common ground among themselves; upstairs in the NHL's executive offices, they knew, there were bigger fights to fight. "Environmental" issues, as panel members called them, were easier, less potentially disruptive—regulations surrounding mouthguards, chinstraps, elbow and shoulder pads, ice conditions and seamless glass—and brought discussion and occasional agreement. Prospective rule changes were harder. They brought out dark suspicions, especially among the NHLPA members. A different game was to the advantage of some players, to the detriment of others, and the NHLPA represented *all* players. Change the rules and maybe good old Tommy, Pierre, and Andrei can't play anymore—even if good new Chris, Jean-Marc, and Igor now can. As teammates in the NHLPA, how would they ever be able to look Tommy, Pierre, and Andrei in the eyes again? The players, who seemingly had the most to gain with such safety related changes, reacted as if they had the most to lose. For them and for the league, this was still primarily a time to concentrate on playing conditions, not life conditions. And the marching orders for everyone were indisputable—no hooking, no holding, no whistles—*keep the flow going.*

Even in the late 1990s and early 2000s when more players were suffering head injuries, each one seemed its own unique case. Mike Richter, from a shot off his mask; Scott Stevens, from one that struck him in the face: they were flukes. Eric Lindros: no matter how big a player is, how many times can you go up the ice with your head down? Paul Kariya: he scores on a shot; his opponent, Gary Suter, is angry and levels him; he was in the wrong place at the wrong time. Stu Grimson, Gino Odjick: just too many fights. And all the others: Brad Werenka, Dean Chynoweth, Paul Comrie, Jeff Beukeboom, Geoff Courtnall, Adam Deadmarsh, Jesse Wallin, and many more—lesser known players. For them was it age, or injury, or maybe they were no longer good enough? One season they were there, then they weren't.

Some players tried to come back; some many times. The public understands knee injuries. They know that stuff can be taken out of a knee, or a knee can be sewn together only so many times. They could see Bobby Orr limp, and probably knew before Orr did that the end of his career was near. It wasn't the same for Pat LaFontaine, or Kariya—not at first—and for them, things never seemed not possible. They looked healthy, they sounded good. There was no outward indication of their injuries. They had symptoms, but then, as if overnight, as with Keith Primeau, their symptoms would disappear. If injuries can be experienced in mysterious ways, why not recoveries? These head-injured players seemed to stutter, then fade into retirement. There was no dramatic, career-ending moment; nothing to bring attention to them or to their injuries. Many of them retired months, even years, after the public had assumed they were already gone.

Even in football, head injuries had long seemed individual and isolated. NFL star quarterbacks—Steve Young of the San Francisco 49ers in 1999; Troy Aikman of the Dallas Cowboys in 2001—retired because of head injuries. A few scientific studies had been conducted at the time and made public; articles were written; stories came out about the 2002 demise of former Pittsburgh Steelers' centre Mike Webster. There were even a few suicides, but these were chalked up more to players' mental issues than to injuries sustained on the field, so the attention paid to the issue came and went. But then there were more studies, and more articles. There were congressional hearings and conferences. More than that, every day there were the games themselves. In every NFL and major college game, every NHL game, if a hit was big enough it was played across the continent on the "Top Ten" plays/misplays/hits of the day. Stuff that made viewers moan, and laugh, and cringe, and groan—often at the same time. Stuff that to viewers eventually didn't seem like one-offs, that came to seem connected to other injuries and other tragic stories—of players they had once followed and loved—to injuries and tragedies of athletes from other sports, even to stories of servicemen returning from Afghanistan and Iraq with PTSD. And all of it, in time, came to seem connected crucially, frighteningly, to their own kids. Around 2009, the head injury pieces began to come together. Concussions had become a major problem in sports.

On January 1, 2011, during the NHL's Winter Classic at Heinz Field in Pittsburgh, Sidney Crosby was hit by David Steckel of the Washington Capitals and received a concussion. Crosby had been circling in the Capitals zone, and had turned his head to follow the puck as it was going back up the ice; Steckel, moving after it, clipped Crosby's jaw with his shoulder. The next game, Crosby was hit again, much less hard, but then he didn't play again for ten months—until late November the following season. Before Crosby's injury, the increased number of players with concussions had been described in terms suggesting an epidemic, or bad luck, something that had come out of nowhere and would disappear on its own just as quickly; certainly not reflective of how the game is played. Crosby's injury and his rehab were monitored by fans and media almost as intently as his goals and assists had been. At first, there was routine concern. He might be out for a few games, or maybe longer; these injuries are unpredictable. Then days turned to weeks, and then months. What was important now was that he be ready for the playoffs, it was suggested, just as it had been for Primeau. Then the playoffs came and went. Even better, some said, now Crosby has the whole summer to get healthy, to go into the next season at the same starting point as his teammates and the rest of the league. Then the beginning of the next season came and went. Concern, which had turned to hope, turned to fear. Crosby's career, or at least his career as a *superstar,* might be over.

The Crosby injury hit the public hard. Not just because Crosby was so good and so young—he was twenty-three at the time—but because he played the game the way it is supposed to be played: hard, physical, with the puck, on the puck, never shying away from the corners or the front of the net, driven to score and to win. He wasn't small, he didn't play with his head down, he rarely fought; his injuries came as a result of normal plays, not flukes. All of the individual, one-off explanations didn't apply. If a player like Crosby was at risk, the game was at risk.

And for parents, if Crosby was at risk, their own kids were at risk. A hockey life is a full family life. It takes a family's time, spends a family's money, monopolizes a family's priorities and emotional energy. It offers the kids an experience with other kids, with a team, with goals and dreams, ups and downs, that is never forgotten. For parents of earlier generations, the decision for their kids to play sports was easy. The rewards were not as great, the commitment required was much less—and besides, there wasn't much else for their kids to do. Now there were a lot of other things to do. There was a choice—other activities, that involved less team-time and more family-time, that were less dangerous. Give parents something to worry about and they will worry. They know their kids might get hurt doing whatever they are doing, and might get hurt more often in sports. But this was different. Head injuries carry such unknowns. A leg that limps is one thing; a brain that limps is another.

So why not play soccer, many parents thought. It's so healthy. The kids just run around, and all you need are shorts and a ball. Or basketball? Or skiing, as Paul Montador had hoped for his kids—something they could do as a family; spend their weekends doing together. Concussions gave parents looking for a way out, a way out. And beyond the tens of thousands of individual families suddenly presented with a choice—what did it all mean for the game itself? Who was going to play, and who wasn't? Hockey's first game was played eight years after Canada was born. It has been part of the way Canadians live since then. What happens if Canada's collective experience is no longer so collective?

As Canadians wrestled with this awkward reality, Gary Bettman was wrestling with his own: what to do with all these concussions, all this attention, all this distraction from all the great things that were happening in the game. He turned defensive. He retreated into his lawyerly instincts—*where is the evidence; I don't accept the premise; prove it*—his hockey guys retreated into their lifelong instincts—*this is a tough game; this is how it's played; this is hockey.* They looked at the video evidence of each on-ice "event" rigorously. Each was the result of an individual incident—the guy ducked down here, elevated himself there, intended this, targeted that. Each involved an individual player with his own individual history—a repeat offender (or not). Each had its own individual explanation. Each was a one-off. The punishment for each was decided with careful consistency. Every decision followed the logic of every decision before—*this one-off was a little more (or less), better (or worse) than that one-off.* Each focused on the trees and missed the forest.

It was like two different worlds. In one world, crushing hits nit-picked to death by the game's decision-makers and commentators—*was his head down? a little, or not;* a two-game suspension, maybe three. In the other, stories were written chronicling the deterioration or death of players from the recent past. The two worlds weren't connected. They couldn't be connected from Gary Bettman's perspective, because connection would mean liability, and liability would mean money and industry uncertainty. Besides, who said they were connected? What science, as determined by what court of law?

The head injury story in hockey is long and winding. Once, head injuries were rare, the result of accident or outrageous actions by players; now they are more frequent. Once, their consequences seemed immediate and certain; now they are both immediate and long-term, certain and uncertain. Once, decision-makers were unaware of the science of head injuries, then they paid little attention to it, then disputed it, and then accepted its role as paramount, pointing out that they will follow that science—*when science knows.* It is the same strategy employed in every industry under siege—tobacco, lead, asbestos, coal, oil—whether the issue is lungs, heart, nervous system, brain, or climate change. There is no need for those in charge to prove anything. They need only create doubt. Besides, smokers, coal miners, hockey players—they know the risks. For decision-makers, it is an issue to be managed, not a problem to be solved.

"Not only must justice be done; it must also be seen to be done." Yes. But not only must justice be seen to be done; it must also be done.

The historic compromise in hockey between performance and safety has gone out of whack.

CHAPTER FIFTEEN

Steve missed the first eight games of the Panthers' 2007–08 season after having arthroscopic surgery on his knee during training camp. The team got off to another uninspiring start, and didn't reach .500 until mid December, stayed there for two weeks, then didn't reach the same level again until March. The team allowed thirty-one fewer goals than they did the season before, but also scored thirty-one fewer and never contended for a playoff spot.

Steve's contract with Florida was up at the end of the season, and he was now a free agent. The Panthers had missed the playoffs for the seventh straight year. The team had changed coaches, made trades, signed veterans, waited for young players to improve, but the Panthers simply weren't good enough. They needed some players that would make them better. Steve could play with better players. There were players like him even on Stanley Cup–contending teams, but he couldn't make the Panthers better in the way they needed to improve. He also wanted to play on a contending team himself—he had had the taste of that in Calgary. The Anaheim Ducks offered him that chance, signing him to a one-year contract for $800,000, the same amount he had received in Florida. And California was California— flip-flops, shorts, and cafés on the beach.

The Ducks had won the Stanley Cup in 2006–07, and lost in the first round of the playoffs a year later. They had two young stars, Ryan Getzlaf and Corey Perry; three ageless wonders in Teemu Selänne, Scott Niedermayer, and Chris Pronger; and a big, tough supporting cast, led by George Parros, that had topped the league in total penalty minutes and fights the previous two seasons. "We were a rough and tumble team with no shortage of guys who could play and also drop their gloves," Parros says. Steve was signed to help make the Ducks even more of what they already were and wanted to be. He dressed every game, played regular 5–6 defenceman time (over sixteen minutes a game), was second on the team in plus-minus, third in blocked shots, eighth in hits, and second in fights, with eleven (his last two with the Bruins) behind Parros—the highest total of his career.

The Ducks drifted back into the middle of the pack that season, yet they always seemed contenders. Their big names were still there. But their checking line—Sammy Påhlsson, Rob Niedermayer, and Travis Moen—so important to their Cup victory, was gone. In a salary-cap league, lesser players on a winning team can always find bigger contracts elsewhere. As well, the Ducks didn't *need* to win as much as they had in 2007. The Kings had been L.A.'s team; they were in the first wave of NHL expansion in 1967; they had had Gretzky. But in 2007, the Ducks were the first California team to win the Cup—their Stanley Cup rings even said so. But having won so recently, they could lose and the sky wouldn't fall. At the trade deadline and going nowhere, the Ducks sent Steve to Boston.

Steve reacted as he always did: he was disappointed, then he was excited. The Bruins were an Original Six team. They played a tough, lunch-bucket style, and had the best record in the Eastern Conference. At the deadline, the Bruins also added veteran scorer Mark Recchi; they were gearing up for a Cup run. Two weeks after the trade, in a game in Toronto, Luke Schenn of the Leafs hit Stéphane Yelle into the glass from behind; Steve took on Schenn. The next night, in Philadelphia, the Flyers scored early in the game; a short time later, right from a faceoff, Steve dropped his gloves and fought Daniel Carcillo. He lost both fights, but in doing so reinforced his reputation with his new teammates. If the team needed something, even if it was a tough, dangerous, and potentially humiliating task, Steve was willing.

He fought against Schenn and Carcillo in the way he'd always fought. He had good balance on his skates; he knew what he was doing. He gripped his opponents close and tight, and ducked his head and angled it away from the worst blows. And if the linesmen had stepped in to end each fight a few seconds earlier, the result would have been mostly even. But Steve was not a finisher. If the linesmen didn't step in, the tide would, and eventually did, turn towards the better fighter and against Steve. He lost most of his fights, often clearly, but almost never badly, and almost never with a knockdown blow. His opponents, frequently the other teams' heavyweights, knew that Steve wasn't a risk to them. Nor did he represent any proud new notch on their guns. Steve's job wasn't to embarrass them or even to beat them, but to remind them that what they or their team had done wasn't acceptable and couldn't be left to stand. In turn, these heavyweights seemed intent only on winning the fight, not on humiliating Steve. To them, what he was doing was honourable, and they would treat honour with honour. When his fights with Schenn and Carcillo ended, the players on the

Leafs and Flyers benches slammed their sticks against the boards in triumph. The Bruins players slammed their sticks against the boards in respect.

Boston finished the regular season second in the league in points, second in goals for, first in goals against. But in the first round of the playoffs, they had to play the Montreal Canadiens. Montreal and Boston had played each other in thirty-one series in their playoff history, the Canadiens winning twenty-four times— eighteen straight between 1946 and 1987, three straight since 2000. But as it turned out, the series wasn't even close. The Bruins won four straight, outscoring the Canadiens 17 to 6. Steve was plus-2 and averaged over eighteen minutes of ice time a game, the fourth highest among the Bruins defencemen.

Next was Carolina. Boston had finished nineteen points ahead of the Hurricanes during the regular season. The series went to seven games and was decided in overtime, with Carolina winning. Steve played even more against the Hurricanes than he had against the Canadiens, averaging over twenty minutes a game, third most among the Bruins defencemen. He was a plus-3, the third best on the team. But for the Bruins, who had felt they could win it all, the defeat was hard. Less than two months later, on July 1, 2009, Steve became a free agent again.

He would soon be thirty years old. For a good team, when a season is over and before free agency begins, a player like Steve seems a replaceable piece. Somebody is ready to come up from the minors, or might be; by next October, somebody else will be one year more prepared, or could be. Steve was the kind of player that teams add at the end of a season, when the somebodies they have counted on haven't panned out, and playoff possibilities and fears are in their headlights.

The Buffalo Sabres were looking for something that Steve was, not what he wasn't, and signed him to a two-year contract for $3.1 million, which per season worked out at almost twice what he had received under his previous contract. The Sabres had made the playoffs regularly in the 1990s, during the Dominik Hašek years, then missed for three seasons, then made it to the Conference Finals in 2006 and 2007, riding the goaltending of Ryan Miller. But the glow around the team was beginning to fade: ownership was unsettled, Daniel Brière had gone to the Flyers, Chris Drury to the Rangers, and Miller's future in Buffalo seemed unclear. Moreover, the economy of western New York was depressed. Even the NFL's Bills seemed in jeopardy. Hope for the Sabres' future had resided in a very small cohort of potential new owners, big-moneyed western New Yorkers who had made their fortunes elsewhere but who believed in western New York, and who could buy their way into the NHL and have big local impact at low-end Buffalo prices. First had been early cable-entrepreneur John Rigas, then Paychex founder Tom Golisano. At this moment, the Golisano years were winding down and natural gas billionaire (and now also Bills owner) Terry Pegula was not yet on the horizon. The Sabres seemed a franchise with its best days behind it.

"[Steve's] got incredible competitiveness, toughness; he's gritty," Sabres general manager Darcy Regier said at a media briefing when Steve's signing was announced, "and there's still upside in his game." Regier then talked about how the Sabres had sixteen young defencemen in their development camp, all of whom had been drafted by the team—and how with three veterans with their contracts expiring a year from now, Steve could be a good transition player.

———

Before Sabres training camp began, Steve made his annual pilgrimage to Gisele. He went with Steve Valiquette, as he usually did now; Gisele was Vally's life coach as well. She had helped him to have a ten-year pro career, Vally believed, one that had been far from inevitable after he had graduated out of junior in Erie—six-foot-six goalies being seen as far more awkward than ideal at the time. Vally had played in the ECHL in Dayton, Hampton Roads, and Trenton, then was a regular in the AHL in Bridgeport and in Hartford, and in the NHL with the New York Rangers. He had also played in the Kontinental Hockey League (KHL) during the lockout, on the Lokamotiv team from Yaroslavl, Russia, that five years later would be wiped out in a plane crash. Vally was the goalie equivalent of a 5–6 defenceman: an NHL backup, or sometimes the third guy in a team's system. With the Rangers, Henrik Lundqvist, one of the league's best and most durable goalies, played almost every game.

Their visits to Gisele also gave Steve and Vally some time together. The hour-and-a-half drive from Vally's place to Gisele's in Lenox, Massachusetts; the nights at the hotel at the end of each day's session; their ride back and any extra time they could steal. It was a chance to catch up, to get news of each other's family, to hear about former teammates, to go "deep and serious" sometimes, as Vally puts it, but mostly to recall old stories and tell outrageous new ones. One of them only had to say something and they were off, laughing, spilling out their souls, and confirming what they most were looking for: that Monty was still Monty, and Vally still Vally.

One year, anxious to start on their drive, they found themselves in Northampton, Massachusetts, ahead of schedule. They decided to stop for lunch, but when lunch was done they were still too early. Walking back to Vally's car, Steve saw a sign on a storefront: "Psychic," it read.

The rest followed. "Vally, Vally, we gotta check this out," Steve said. Two suckers, about to be born. "So we go in," Vally recalls with a laugh. "It's a tarot-reading card place and the lady's name is Dorina." Dorina began with Steve. They went into another room; Vally waited. Then it was Vally's turn. When they were both done, Steve and Vally left. "Monty's just bouncing," Vally says. "'Vally, Vally, oh my god, she knew about everything. She just nailed it.' And I'm saying, 'Yeah, yeah, but Monty, how do you think she knew? What did you tell her?' And he says, 'No, no, it was amazing. It was the best.'

"Dorina tells him that he's got this great career ahead of him, that he's gonna play in the All-Star Game. Meanwhile, she tells me I'm never going to make the NHL, I'm gonna be a construction worker."

Dorina had seen the future: two idiots coming through her door.

When they were walking towards Vally's car, "Monty was still carrying on, [saying] 'I'm keeping in touch with her. She's going to light candles for me,' and we're laughing. And I say to him, 'Why am I even going to Gisele's? I'm going to be a construction worker. What's the point?' And we laugh. Then Monty starts saying, 'Dorina, you're in. Gisele, you're out.' Now we're both into it, 'Dorina, you're in. Gisele, you're out.'"

After that season began, Steve and Vally were talking on the phone. "I say, 'Monty did you ever call Dorina back?' And he says, 'Oh yeah, she's been lighting candles for me.' And I'm going, 'Come on man,' and he says, 'Yeah, I just cut her a cheque for $1,200.'" Even Vally was stunned.

For the rest of Steve's life, in a phone call or text, or when they were together, out of nowhere Vally or Steve would say, "Dorina, you're in," and the other would say, "Gisele, you're out," and they would crack up.

"Gisele helped us to believe in ourselves," Vally says. "She gave us these little things to help us in this very busy game, when your mind can get very, very busy. We each had our mantra. Before a game, instead of everything being chaotic, everything was calm. Our mantra was the same: 'I now command my unconscious mind to know with absolute certainty that I have all the skills, resources, and abilities necessary to know that I am an NHL hockey player, and I belong here, and I am ready.'" By having Steve and Vally focus on their mantra, Gisele had them *not* focusing on all the other things that might undo them. And when Steve and Vally were done saying their mantras, wherever they were, they would do the same thing. "I would look in the mirror," Vally says, "see my reflection, and say, 'I am ready. I am ready. Let's go. I am ready.'"

"I was cut from teams many times through youth hockey," Vally continues, "and didn't make AAA until I was fourteen. I've been told I'm not good enough a lot. So when I did get good enough, you still have to believe you have a right to be there. That you belong. Later, when Monty got that big contract from Chicago, he didn't believe he deserved it. When you don't believe you're good enough, you never lose the feeling."

Vally also helped Steve believe in himself, and Steve did the same for Vally. One year, when Steve was with the Panthers and Vally with the Rangers, they played against each other in Florida. "I was in the last year of my contract with the Rangers," Vally recalls. "Henrik [Lundqvist] had played the first fifteen games of the season. I played my first game in Toronto, then he played five more. Now it's my second game, in Florida, and I still haven't gotten my letter from the team saying I can move out of the hotel and into my own place. But I'm going after my dream. I want to become a full-time NHL player for the first time, and I'm twenty-eight years old," he says. "And I had a great game." The Rangers were outshot by Florida, and lost in a shootout. "I just had one of those nights," Vally says. When you're a 5–6 goalie, every game is a big game. "After it was over, Monty comes up to me in the hallway, and we're both just sweating. It was like we couldn't wait to get out of our locker rooms to see each other. And he's like, 'Vally, Vally, you're the best goalie that's come in here all year,' and he's as happy for me as you could ever hope a friend or teammate could be."

Steve and Vally talked so often and so long, that at times when Vally saw Steve's name on caller ID, he wouldn't pick up. "Sometimes I didn't have an hour. I knew I wouldn't be able to get off [the phone], and I wouldn't want to get off." Vally stops himself. "I wish I had that opportunity now, even if just for a minute."

Steve gave the Sabres what they expected of him. He played seventy-eight of the team's eighty-two games in 2009–10, averaging over seventeen minutes of ice time a game; he played seventy-three games and averaged almost twenty minutes of ice time the following season. He had bad stretches: when the team swooned late in January the first year, Sabres coach Lindy Ruff, impatient with some of Steve's on-ice decisions, met with him, made him a healthy scratch, then moved him up to the wing before restoring him back to defence. He had good stretches: in mid-November the second year he was leading the entire NHL in plus-minus (plus-13), and was tied for second (plus-16) a month later; this on a Sabres team that was playing below .500. He also had long solid stretches of little statistical note, but which mattered to some of his teammates in different ways.

Mike Weber was a twenty-three-year-old, tough, stay-at-home defenceman for the Sabres, who had played a scattering of games with the team the previous two seasons but had yet to become a regular. "I was in and out of the lineup," Weber recalls, "and [Monty] would sit there for a whole plane ride sometimes, and just talk to me, about his situation, where he came from, and the type of things he had to work through to make it to be a regular in this league." He was a "calming voice," Weber says—someone who "[no matter] what the problem was, what the issue was, no matter what was going on with him, he always had time for every-one." Steve was a teammate. Teammates help other teammates because that's what makes the team better; because that's what teammates do.

But Steve had reached a new stage in his career. He turned thirty-one during his second season with the Sabres. An improving team might improve right past him, and leave no place for him. It would also leave Steve more vulnerable to being replaced by a big-ger, younger, cheaper, much less banged-up version of himself—like Weber. "[Steve] will probably go down as one of the best teammates I've ever had," Weber says. Weber played with the Sabres for the next several years.

The Sabres won the Northeast Division title in 2009–10, and finished third the following year. Both seasons, they were elimi-nated in the first round of the playoffs, first by Boston, then by Philadelphia. In his second year with Buffalo, Steve started strongly but his play tapered off, and in the seventh game of the Flyers' series—in the biggest game of the season—he was a healthy scratch. The Sabres haven't made the playoffs since.

Steve would become a free agent once more on July 1, 2011. He was thirty-one years old. He had been a transition player for the Sabres—helping them to get better and to allow time for younger defencemen to develop. Some of those younger blueliners were now ready for more; and others, three months before the next season began, might be by the time October came around.

A player wants to be a free agent during the prime years of his career. Then he has teams competing for his services. He knows himself, and what he values most. He can chart his own path, and go where he wants to go. To play for a winner? To play with the pressure and expectation of winning—or not? To play in what city, with what climate, have what lifestyle? But outside a player's prime years, free agency is a curse. They become just another resumé in a very high, desperate stack on a GM's desk. Even being on the wrong team under the wrong contract is better than having no team and no contract at all. That prime/post-prime moment, where a player passes from chased to chaser, from stuck-with-them to free-of-you, happens as if overnight.

In March, Steve had written the names of five teams in his journal: New Jersey, Boston, Toronto, Buffalo, and the New York Rangers. He also wrote Philadelphia, and crossed it out. He created a decision grid for each team, drawing a large rectangle on the page that he divided into four smaller ones: *What will I/won't I get if I sign with __? What will I/won't I get if I don't sign with __?* The criteria that he mentioned most often were: ability to contend; travel (Eastern Conference); proximity to home; and sunshine. Beside New York, Boston, and Toronto, he also wrote: "great city."

On June 29, 2011, the Sabres traded Steve's rights to Chicago for a seventh-round pick in either 2012 or 2013. This gave the Blackhawks two days, exclusively, to sign him to a new contract. The next day, June 30, Steve's agent, Kurt Overhardt, called him with the news. Kevin Magnuson, the son of former Blackhawks captain and coach Keith Magnuson, was an agent in Overhardt's office. He heard Steve on the phone's speaker system. "I'm so emotional right now,"

Magnuson remembers him saying. "I can't believe I just signed this deal. It's amazing. Thank you so much."

"He was almost crying over the phone," Magnuson recalls. The contract was for $11 million over four years.

The $11 million was a big deal for Steve. The fact that it was with the Blackhawks, who had won a Stanley Cup a year earlier and had the ambition and the maturing young stars to do so again, made it an even bigger deal. But what made it the biggest deal of all—why Steve had almost been crying on the phone—was that for all of his career he had been playing on one-year or two-year contracts, and for much lesser teams. This was the Chicago Blackhawks, and this was for *four* years. They were committing to *him*. They *believed* in him, in what he was and in what he had become. He wasn't just another 5–6 (or 7) defenceman who filled a spot on a team until someone better, younger, and cheaper came along. He might still be a 5–6 defenceman, but now he was one that was being counted on, even needed, who would be there next year *and* for the future. At this contract price, the Blackhawks might even be thinking of him as a 3–4 defenceman, a guy still with an "upside," even in his thirties—what Steve had always seen in himself, and what other teams had hoped for. He would be in Chicago longer than in any other place in his career; he would be thirty-five when his contract expired. This might be his last team. This might become home.

On February 5, 2011, near the end of his second season with the Sabres, Steve suffered a concussion in a collision with the Leafs' Jay Rosehill. Steve returned to the lineup thirteen days and six games later. A few days after that, he wrote in his journal, "Keith Primeau,"

"Play It Cool," and "Sports Legacy Institute." Play It Cool is the head-trauma assistance program that Primeau had co-founded to work with minor hockey organizations to reduce concussions. The Sports Legacy Institute, now the Concussion Legacy Foundation, was co-founded by Dr. Robert Cantu, the noted neurosurgeon and concussion expert at Boston University. A few days earlier, there had been a story in the *Globe and Mail* about Primeau that made reference to the recent discovery that Bob Probert, the former enforcer who had died the summer before and who had donated his brain to Boston University, had CTE.

CHAPTER SIXTEEN

As Steve was about to begin his career in Chicago, Marc Savard was ending his in Boston—and in the NHL.

For fourteen seasons, Savard had made magic. He wasn't big for a hockey player, even as a kid. But he was a good skater, and what set him apart was that he could "see" the game. He always seemed to know what should happen next—and then, with hands as quick and adept as his mind, with a soft pass, he would make it happen. His joy was making plays. Offensive plays, his critics sometimes pointed out, not defensive ones. Twice he was the OHL's scoring champion, playing for the Oshawa Generals, but he was drafted only in the fourth round by the Rangers, the 91st overall pick, because of his size. He spent his first two seasons of professional hockey moving between the Rangers and their farm team in Hartford, then got his chance after being traded to Calgary. For the remaining twelve years of his NHL career—first with the Flames, then the Atlanta Thrashers and the Bruins—he would be one of the top and most reliable scorers on his team.

During his junior years in Oshawa, Savard had discovered Peterborough, only an hour away. In years after, he would work out in the summers with Jay, Nick, Keats, and Steve. He and Steve also overlapped in parts of two seasons with the Flames. His career ended in 2011 after suffering a concussion, his second in ten months, while playing with the Bruins.

Savard remembers the concussions he had as a kid, but now only because he has reason to think of them. At the time, they seemed to be only big hits, moments that teammates would laugh at and re-enact in the dressing room, as he did theirs; tough lessons that were learned the hard way. There was that game in Metcalfe, just south of Ottawa. He was playing Junior B, only fifteen years old—some play-ers in the league were three or four years older. It was early in the season, and he was cutting through the centre of the ice, reaching for a pass, and got blasted. "I remember pretty much blacking out," he says, "then crawling to the bench. I said to my dad after the game, 'Geez, Dad, maybe we made a bad decision here. There are some big fellas in this league.' My dad just said, 'Let's stick with it a bit more and see how it goes.' I don't think I missed any games."

He also recalls an NHL game, with the Flames. "It was after a whistle, there was a little bit of a melee and a guy punched me in the back of the helmet." Savard went to the bench, the trainer checked him out and asked him how he was. He said he was fine— "You're a hockey player. That's what you say"—and played the rest of the game. The next night, in the second period, he suddenly started to feel nauseous and tired. What followed, Savard says, "was like a week of hell." The nausea eventually went away, but the fatigue stayed. "I was exhausted. I slept for three days straight." But then he started to feel better, and after a week he was back. "I just played on, and everything seemed good."

His next concussion came a few years later, when he was playing with Atlanta in a game against Montreal. "The Bell Centre had seamless glass and it was rock hard," he recalls. "[Stéphane] Quintal hit me behind the net and my head hit the glass. I'm not even sure how much I missed that time." In fact, he was out ten days. "But it was the same kind of thing. I was exhausted."

Savard knew his body. If his shoulder felt a certain way, often even before a doctor or trainer told him, he knew that he'd be out of the lineup day to day, or seven-to-ten days, or he would try to play through it and not stop playing at all. He had had several knee injuries. He knew the symptoms and treatment. He knew how his knee would feel each day, and through each stage of healing, and he knew he would get better. Injuries are just a part of the normal rhythm of a season and of a career. They were not, and could not be, his focus. His focus was healing and health. It was getting back in the lineup. It was playing. No matter how severe, an injury is almost never career-ending when it happens. If doctors, procedures, and medications aren't the answer, time and hope might be.

By this time, Savard was beginning to know his head as well as he did his knee. After each damaging hit, he knew he would feel an incredible fatigue that would last a few days, then he would start to feel better, and then he would go back to play.

Savard calls his next concussion the "massive one." And while it represented the turning point of his career and was the cause of great debate in the media and around the league for days afterwards, Savard isn't sure when it happened. "I think it was April, maybe January or February." In fact, it was March 7, 2010. "I just got hit, and I don't recall the hit at all. I was out cold for thirty seconds."

The Bruins were in Pittsburgh; the Penguins were leading 2–1 with less than six minutes remaining in the game. Boston forward

Milan Lucic carried the puck down the left side, crossed the blue line, and passed the puck back into the middle to Savard. Penguins forward Matt Cooke saw the play developing and moved across the ice towards him. Savard took his shot, the puck left his stick, and he was extended into his follow-through and completely blind to Cooke cruising into range. At the moment Savard was at his most defenceless, Cooke, moving several miles an hour, struck Savard's head with the full force of his 200-plus-pound body, making only incidental contact with the rest of Savard. Savard's head snapped around, his body swung out of control to the ice. He lay on his back, motionless except for occasional twitches of a hand or foot, and then was carried off. In his team's dressing room after the game, Cooke explained, "I just finished my check." He was not penalized on the play.

The Bruins flew to Toronto for their next game; Savard remained in Pittsburgh at the team's hotel. The following day he was seen by a doctor, did some tests, and was released, and he flew back to Boston. "I was just feeling frustrated," he recalls, "and irritable. I couldn't wait to get home. When I got there I remember just shutting the door and saying, 'I'm going to bed.' Then I just slept and slept and slept."

He wanted darkness, silence, stillness. He wanted to be alone. So he slept during the day and stayed awake all night. He had always been something of a happy-go-lucky guy, but now people made him irritable, as he says, and he didn't know why; and he didn't like feeling that way, and he didn't like not wanting to have people around him. But people brought light, and movement, and noise, and all of these things suddenly seemed to him so bright and quick and loud. "Everything bothered me," Savard says.

This continued for weeks with little change. "These were some of the toughest days in my life. I just didn't see the light at the end of the tunnel at that point. I felt like crap every day."

He's not sure whether it was because his symptoms diminished or, as he put it, "the hockey player in me started to come out because the playoffs were just around the corner." But he started to get excited; he had something to look forward to. He took the tests he needed to take, and passed them. "I got back on the bike and started working out again. I was getting myself ready for the play-offs. Sometimes, I admit, when they asked me how I felt, I told them great, and told myself great, because I wanted to play."

The Bruins lost the first game of their first-round series against Buffalo; Savard was skating but not yet ready to play. As his return got closer, the Bruins started winning, so Savard rested a little longer. Boston won the series in six games, and in the next round they faced the Flyers. Savard returned to play.

Not playing for two months, jumping back into the fire of the playoffs, Savard didn't know how he would feel. "I don't remember the total feeling I had," he says. "I remember feeling just still tired." It was a different kind of fatigue. It had nothing to do with the state of his conditioning. It was as if his life-energy had been sucked out of him. "Suddenly, I just didn't have anything to draw from." The first game went into overtime; Savard scored the winning goal.

The Bruins won Game 2; Boston coach Claude Julien was carefully monitoring Savard's ice time, limiting him to about fifteen minutes a game. In the third game, Bruins centre David Krejčí broke his wrist in the first period. Krejčí had taken over some of Savard's roles when Savard was injured; now Savard needed to do the same for him. The Bruins won Game 3.

The fourth game went into overtime; Savard played over twenty-four minutes, the most of any Boston forward. The Flyers won, cutting the Bruins' series lead to 3–1. At first, Savard had tried to persuade himself that he was tired because he hadn't played in the previous weeks. Then, he thought, maybe if I keep playing I'll feel better. When Game 4 was over, Savard recalls, "I don't say anything to anybody, but I'm done."

The Flyers won the last three games, and the series. When the round was over, the *Boston Globe* wrote, "Marc Savard didn't have the ending he wanted. In 15:53 of ice time, Savard had three shots, was on the ice for two Philly goals, lost 7 of 10 faceoffs, and was involved in the too-many-men blunder in the third. Clearly, Savard needed more time to recover his touch after missing two months because of a Grade 2 concussion."

"After the series, I went into a bit of a depression," Savard says. He was tired, but he was used to being tired. He was disappointed, but he had been disappointed before. This was something more. "I really felt horrible. I remember thinking, 'I don't know what's going on with me. I don't feel like myself.' I was just feeling down, really down on myself. Down on everything." His girlfriend, now his wife, was in Peterborough. "I basically didn't want to be with her because I was just irritable. I didn't want to talk. I think I was pretty mean at the time."

He went back to Peterborough, took it easy, slept, but nothing changed. When training camp began in September, he was still in Peterborough. Later that month, he drove to Boston and spoke at a press conference. The next day, Kevin Paul Dupont of the *Boston Globe* reported:

A somber Marc Savard, his eyes welling up a couple of times, stood in the TD Garden dressing room yesterday morning and made it clear he won't be able to play hockey for a while.

"I am definitely going to take my time," said the
Bruins' No. 1 center, "and make sure that I am 100 percent
in every aspect before I even think about playing."
What ails Savard, in the broad and often-ambiguous
definition, is postconcussion syndrome.

Dupont then listed the most common effects: "nausea, headaches,
dizziness, seeing spots, and depression." When he asked Savard the
most difficult one for him to deal with, Savard replied, "Oh, probably
the depression part." The article states that "his tone [was] somber,
his emotions clearly stirred."

"With no training camp," Savard recalls, "I was having a tough
time. I didn't know what I really wanted to do. I wasn't really interested in hockey because of what I'd been through." But now he was
in Boston. "I got back to being around the rink, and seeing the guys,
and then I started working out a bit."

He started feeling better again.

On November 19, Savard got the go-ahead to begin practising
with the team. After his first workout, he reported that he felt "great."
Four days later, he went to Pittsburgh to be tested by Dr. Micky
Collins, a well-known clinical psychologist at the University of
Pittsburgh Medical Center. The session lasted six hours. "I went
through tests that I'd never gone through in my life," Savard says.
The clinic's staff pushed him almost to the point of exhaustion, then
at his weakest, they tested him some more. He was also given the
baseline tests that he had done every year at training camp. "In
some areas I passed with flying colours, and in others, I wasn't any
worse than I had been before. They said I was ready to go if I felt
ready to go."

Savard was now cleared to participate in contact drills.

His press conference in September about his depression and recovery had touched a nerve in big, tough Bruins fans. In an interview in late November, Savard spoke of how important the support of those fans had been to his recovery. "[Y]ou find out that so many people have struggled with this thing. All those letters from fans. I got at least [ten] thousand. . . . I was going through a tough time, then to read those stories. . . . It's pretty incredible, huh?"

Since that game in Pittsburgh almost nine months before, Savard had had one wish: "I wanted to feel normal again. I wanted to do normal stuff." So he did all kinds of treatments and took all manner of tests so the dizziness and depression would go away, and mostly they did. Then he realized that the absence of symptoms didn't make him feel normal; only playing did. If his symptoms wouldn't go away, that didn't mean he couldn't play. That meant he *had* to play. This wasn't about feeling normal to play hockey. This was about playing hockey to feel normal. *Normal was playing hockey.*

There was nothing in the tests that said he shouldn't try. It was up to him, and how he felt. "I was on the ice a lot more, and skating and working with the workout coach," Savard recalls, "just getting ready for that day to come again where I could play."

It didn't take long. "I don't remember the date. I just remember the crowd and the standing ovation. It was a really great night for me. I was a little emotional. I missed playing so much. And after that I started to feel pretty good." He was "with the guys again, doing the everyday stuff that I'd done for the last fourteen years. I'm good. I'm good." Three weeks and ten games later, a story in the *Boston Herald* describes him as "a shadow of his former self."

"I wasn't as effective, that's for sure," he says about his play after he returned. "I'm a guy that saw the ice really well, and was always a step ahead." Now it was as if everything was just there as it happened. He wasn't foreseeing anything. A big player can power his way to open ice; everyone else has to get there first. If Savard couldn't do that, his whole game wouldn't work. He'd always have to play in a scrum of bigger, stronger players, with no space or time to see the ice, to make his magic.

To quiet his doubts, he told himself that he hadn't played for the better part of a year. It had only been one bad game, after all, or two, or only one bad week, or two. He just needed to get his feel for the game back, and then he'd be able to see everything better and faster again. He just had to stick with it.

Things did come back in time, Savard remembers, at least here and there. "I definitely had some moments where I felt real good. I had a couple of games that I felt like my old self. Making play after play some nights, then I'd go about three games not doing anything. But then having another good night." He looks back on that time and on himself, and says, "I never thought about this until I stopped playing, but I don't think I had it anymore."

On January 15, 2011, the Bruins were trailing Pittsburgh, 3–2, in the third period in Boston. Savard skated the puck up the right boards across the Penguins' blue line and, with time, turned to face across the ice, waiting for his teammates to get into scoring position. Pittsburgh defenceman Deryk Engelland closed in on him. Engelland is a few inches taller and was skating in a mostly upright position. Savard, bent slightly, made the pass as Engelland, with his hands and arms extended, pushed Savard's head into the glass. Savard went down on his knees, holding his head. The Bruins trainer rushed out. Quickly, Savard was up and skated to the bench, and was back playing on his next shift.

After the game, he said he was "a little woozy." The next day, his neck was sore, but he had no concussion symptoms. He had received a blow to the head, but except for a few short-term side effects, he was fine, he thought. This was a good test. He must be healing.

A week later, the Bruins were playing the Avalanche in Denver. Savard chased down a loose puck in the Colorado zone, almost into the left corner. Avalanche defenceman Matt Hunwick went with him, and as Savard made his pass behind the net, Hunwick—with a short, sharp motion—drove into him, Savard's body hitting the boards, his head hitting the glass. "It wasn't anything massive," Savard recalls. "But I remember falling to my knees and seeing black again. And I'm in a bit of a panic, and I stay there, and when my eyes come back, I could see dots. I remember seeing the trainer coming out and I told him, 'I think I'm done. I just can't do this.'"

What disturbed Savard so much was the sheer routineness of the play, the kind of hit that can happen a few times in any game. To Savard, it now didn't matter what he felt like the next day, whether his symptoms were better or worse than before. It wouldn't matter what doctor he saw, or what tests he passed. "I had battled through it all. I tried three times to play again, and it just wasn't going to happen. It was the last game I played."

It is now several years later. "I feel pretty good," Savard says. "I still have my issues. When I say 'pretty good,' my wife will tell you that I'm not good. I have some anxiety, some panic situations that I get medication for. But I don't get too many headaches anymore. I have some memory issues, especially the short-term. But other than that I'm pretty happy. I help out with hockey, and try to keep myself occupied, and that's a big thing for me."

He had never had panic attacks before. "I ended up going to the emergency twice a few years ago. I thought I was having a heart attack. They tested me and told me my heart was fine." He skates occasionally, but takes that pretty easy. He plays golf all summer and is still good enough to play in some Canadian Tour events. He has started working out again with a trainer. "But it's just thirty minutes a day," he says. "Nothing strenuous. Just to keep me active." As for any kind of nine-to-five job, "That would be impossible, I think."

He and his wife saw the movie *Concussion* shortly after it came out. "It was quite moving for us to watch. She was really disturbed," he says. "It kind of scared me seeing what happened to some of these guys later on in life, and their anger issues, and what they were doing. I was basically telling myself that they had taken way more blows to the head than I ever did. It was the only way I could say to myself that it's not going to happen to me, right?

"I just try to keep everything normal and to live a normal life. So that's kind of what I've been doing." Hockey was always at the centre of normal. Now he has a new normal: "Going to the kids' hockey, raising a little girl, playing golf, spending time with my wife, and just trying to be happy."

"It was sure a difficult time in my life," he says, thinking back. "It's not something that I've ever really talked about. I try to stay out of the spotlight. But I guess stuff needs to be said at some point."

CHAPTER SEVENTEEN

Before Chicago's 2011 training camp opened, and before Steve and Vally's annual "head" session with Gisele, Steve had his yearly "spirit" week in Vail, Colorado. The idea had started from a conversation between Marty Gélinas and Andy O'Brien. NHL players were training harder than ever in the off-season, but they were finding that after they arrived back in their NHL cities and before camp began, the energy of their workouts dropped. It was the last gasp of their summer, so it seemed they had a right to take it easier. There were also so many players of so many different skill levels in the on-ice sessions that it was impossible to get much done. The players, Gélinas and O'Brien realized, were losing fitness at the moment that they needed it most. Gélinas wondered, "What if we were to go away for a few days and have high-intensity workouts to get us ready for the season?" So he and O'Brien approached Ed Belfour, who was with Florida at the time; Belfour was all for it, and suggested Vail as the site. He had been there with the Dallas Stars; Vail was beautiful, and wasn't busy at that time of year.

There were about eight or nine players in their initial group, Hayley Wickenheiser and Steve among them. As word spread, in subsequent years the numbers grew to twenty-four. It was by invitation only, and it was an invitation highly prized because of O'Brien and his work with Sidney Crosby—and because Crosby was there, too. How often, even as an NHL player, do you have a chance to spend five days with the world's best player? To share the same locker room, the same ice surface, the same gym and breakfast table? To see what makes him tick? And because Crosby was there, John Tavares was there, and Tyler Seguin, and Jason Spezza, and other guys that they themselves wanted to be there—good players, hard workers, great guys. It was hockey's Davos.

The first night, the players picked the teams, then went at each other the rest of the week. Golf, basketball, mountain climbing, bowling, three-on-three hockey; Team Crosby against Team Tavares, the players pushing themselves and each other the way no coach can; no standings in the newspaper, no money on the table, but stakes that couldn't be higher—respect and bragging rights. It was the ultimate street hockey game. And if any of them thought of some other game to play, they competed at that, too.

"It's at more than eight thousand feet of altitude," O'Brien explains, "so it's really difficult. But it's this incredible environment removed from all the distractions of your life, where you're just able to focus on what you are doing, whether it's fishing, or golfing, or climbing a mountain to see a lake that's very pristine and hard to get to. And Monty was at the centre of it. He planned the dinners. He planned the golf and put the foursomes together. He kept score of every contest, because at the end of the week a trophy was presented to the winner.

A few years later, after the camp began, Monty approached

Vally, who had just retired. "'Hey chum,' he says to me," Vally recalls, "'you could be the goalie coach for our camp.' And I'm thinking, I'm five minutes retired and I completely don't belong. Then Monty goes to Andy [O'Brien], 'Andy, we've got to bring in my boy, Vally. We need him at this camp. He's a character. He's going to be a guy that can glue everybody together.' And Andy says, 'We don't really need a goalie coach, but okay Monty.' And they bring me on." Vally was amazed by what he saw. "These players have a different relationship than anything I ever experienced in the NHL," he said. "They are stars for a reason. Crosby versus Tavares in a faceoff circle; it was just awesome."

Awesome, too, was Steve. "I just loved seeing him chirp Sidney Crosby. I loved seeing him chirp, absolutely chirp, John Tavares. These are the stars of our league, that we all look up to, and they love him! Sidney Crosby loved Monty. They were workout partners. Monty was texting him. How did Steve Montador, my buddy, and just a regular NHL dude . . . ?" Vally answers his own question: "Because he treated the stars like they were just regular fourth-liners. You know, like any guy is walking into the locker room and there's Monty, 'Nice pants, chum. You gonna change before we go out?' Next guy, 'Not a bad shirt. I wouldn't wear it to a shit fight.' You can't say that to Sidney Crosby. And Crosby's loving it! It was priceless."

"It was like Monty was the superstar," O'Brien says. "It was really funny to see Sidney Crosby taking a liking to him, following him. Monty had this ability to walk into a room and be *the guy*."

O'Brien remembers a moment the first year they were there. "Eddie Belfour had a friend who had made some money in the oil industry and had this beautiful home in one of the most expensive areas in all of the Rocky Mountains. We went up there just for a little meet-and-greet and to enjoy some time together as a group.

And Monty starts talking economics and investments and real estate with this guy." O'Brien shakes his head as he tells the story. "Monty just morphed into being whatever he needed to be in any situation. He took control of the room. Not because he wanted to. That's just who he was."

O'Brien also remembers a hike they all went on. "It was a jog-hike, up the mountain, a little above ten thousand feet, and then around a lake, so the guys could walk for a bit, then run again. It was probably an hour long. We had already trained an hour that morning; Monty about ninety minutes. And we're on this trail and Monty has his bag with him with a camera in it and every kind of power gel and supplement you can imagine. Everyone else is scared about a hike at this altitude, and here's Monty carrying a backpack that weighs twenty pounds. And he's at the front of the group. Then he sees a bear." O'Brien laughs at the memory, and talks even faster. "So he cuts off the path, away from the group to take pictures of this bear! He's gone. Then about ten minutes later, we see him, and now he's at the back of the group, but then he catches everybody and ends up way ahead again. These are some of the fitter guys in the league." O'Brien pauses a moment. "It was actually one of the few times in my career where I took a step back and said, 'Wow, that was very impressive.' He was a cardiovascular machine."

"Monty was just such a unique human being," says Vally. "It's hard to even explain. Actually, it's not hard to explain, but it's hard to believe. I just loved watching him at Vail, seeing him absolutely mature from being captain of the rookies to captain of the stars."

During Steve's first few days in Chicago, he ran into Daniel Carcillo, a tough, feisty left winger, who was also in his first season with the Blackhawks. They had met twice before, the first time years earlier when they were taping segments for the TSN show *Off the Record*, with Michael Landsberg. "I was twenty-two years old then," Carcillo recalls. "I remember I'd had two or three beers downstairs because I was really nervous and wanted to loosen up a bit, and when I went upstairs there was this guy getting his makeup done. He had cool hair and tattoos everywhere, and I wondered who he was. He looked me right in the eye all the time he was talking to me. I will never forget meeting Monty for the first time. He had something I wanted. I knew that right away. He was happy and he seemed so comfortable with himself, so sure of himself." The second time they met was when Carcillo was in Philadelphia and Steve in Boston, when they fought. That was in 2009.

Carcillo had modest size and skills, but he played with the ferocity of someone who had a grudge against the world and whose survival was at stake. "Car Bomb," he was called. He had made a name for himself leading the NHL in penalty minutes with the Coyotes in 2007–08, but made his breakthrough with the Flyers in 2009–10, when the team went to the Stanley Cup Final and lost to Chicago. The next season was worse for both the Flyers and for Carcillo, and when it was over, his contract up, the team didn't extend him a qualifying offer. At age twenty-six, Carcillo was without a job, and with a reputation on and off the ice. There is good "out of control," when a player unnerves and unsettles an opponent, and bad "out of control"—on the ice and off—when he unnerves and unsettles his own team as well. Car Bomb had become a time bomb.

On July 1, 2011, he signed a free agent contract with Chicago. The Blackhawks, after winning the Stanley Cup in 2010, had been pushed around in an opening-round loss to Vancouver the next season. A headline in *USA Today* announcing Carcillo's signing read: "Daniel Carcillo is ready to stir it up in Chicago."

Off the ice, Carcillo was less sure of himself. "I was going through a really rough time in my life. I was lonely and not in the right frame of mind." When the Flyers let him go, he had made a decision to get sober and turn his life around. He decided he would do it by himself, the way he liked to do everything.

Early in his career with the Coyotes, Carcillo had been charged with DUI, which got him the attention of the league and the NHLPA, and he was put into their substance abuse program. "I fought them for about five or six years," he says, "asking them to leave me alone, telling them that none of this stuff applies to me, blah blah blah. And the doctors were really good. They just let me come to my own realization, because if you push a guy like me into a corner, I'm going to fight back." But there was something else Carcillo knew, that the doctors didn't know. "They had no idea what I was doing as far as taking pills, and how deep into that I was. When I blew my knee out and had a few surgeries in Philly, I got hooked on painkillers, and that's what brought me down pretty quickly the last year I was there. When I didn't get signed, that was a wakeup call."

With his new contract in Chicago, he decided to get help. "I went to see Dr. [Brian] Shaw and told him what I'd been doing." Shaw, a clinical psychologist in Toronto, was co-director of the NHL/NHLPA Substance Abuse and Behavioral Health Program. "I started going to AA meetings that day," Carcillo says. "Dr. Shaw asked me if I wanted to go to rehab, and I said no because I only had a month to train [before camp] and I wanted to be around my family. So I sat on my couch for a week just shivering, detoxed on my own, then started working out. That's how I got sober." Then he drove to the Chicago camp. "I wanted to try to do it on my own," he explains, "because I wanted to feel what it's like to come off it so I wouldn't do it again." He also wanted to feel proud of himself for something.

That's when he ran into Steve. For Carcillo, it was three times lucky. "When I got to Chicago," he recalls, "it just so happened there was a guy who had been doing what I wanted to do, and he was seven years sober. But he was just so normal. He wasn't awkward. He was always happy. Confident of himself. Comfortable in his own skin. Well dressed. He just had something different about him. Other people try so fucking hard to be cool; it was as if he wasn't trying. He seemed like he had it all figured out.

"It wasn't like he began to coach me or teach me. I was just following his lead. He was showing me how to live this new lifestyle and to leave that other stuff behind me. People don't decide to make changes unless they've hit rock bottom. I wasn't quite there yet, but I needed to be sober and here was this amazing guy."

They lived about a kilometre from each other. "If I wasn't at his place, we were at a restaurant for dinner, or he was at mine, listening to records, chilling out and talking." They went to concerts together—The Lumineers, The Tragically Hip. (The Hip played "Bobcaygeon," Carcillo recalls, about a little town near Peterborough.) Restaurants and clubs aren't easy places for an alcoholic, but to Carcillo this wasn't about hiding away in a basement to avoid temptation. This was about keeping himself so busy that he didn't think about temptation. It was about exchanging one fixation for another. "When I first got sober," he says, "I kept saying 'Never again. Never again.' But that meant alcohol and drugs were *always* on my mind, and I was giving them even *more* power over me. So staying at home just wasn't good for me."

Sometimes he and Steve went to AA meetings together. "When you first get sober, you do ninety meetings in ninety days," Carcillo says. "I really liked going to the Mustard Seed, which was mostly an all African American meeting. It was really spiritual; almost like church." On Tuesday nights, they sometimes went to a men's meeting. "It was just a lot easier to relate to them," Carcillo says, "and girls can sometimes be a distraction." There might be two hundred people at some meetings; in others, only four or five. "We also went to meditation meetings where the first ten, fifteen minutes were just silence, where you're just thinking, then reflecting on what you're thinking." Carcillo didn't say much in these sessions. Not at the beginning. "I just wanted to listen," he said. "But sometimes they picked you and you had to say something. I just said all the stuff I heard from Monty or in some other meeting."

AA meetings are supposed to be anonymous, but this was Chicago, and every night people came up to him and Steve and wanted to talk about the Blackhawks. "That's why I ultimately stopped going to meetings," Carcillo says. Instead, he and Steve started to see their almost-nightly dinners as AA meetings of their own. There they didn't need to tell their stories, they didn't need to feel the reinforcement of the group, they didn't need to commit and account to others, they didn't need to share their hopes and dreams. "Monty and I were already on the same page with everything. We were working towards being happy without any drugs or alcohol; learning how to live normally." Carcillo pauses as he remembers Steve.

"I think about the feelings that I had while I was with Monty. I felt just so safe. Growing up as a kid, I didn't always feel safe. Monty was so different from everything I'd experienced up to that point. It was intoxicating. This was my new drug."

Carcillo and Steve were a good match: one needed to find a new way; the other had found it. One needed help; the other needed to help in order to help himself.

Steve kept a journal. Before his first season with Chicago began, he wrote:

Season Goals
10 goals, 22 assists, +15, all 82 games, 20 minutes/night
    towards end goal of a Stanley Cup
Value me and I feel valued . . .
Character, consistent, strong physical presence

He also wrote about other goals he had for the year:

Travelling
Photography
Broadening my perspective
Courses—yoga, cooking, piano, guitar, book clubs,
        movies, AA, volunteer organizations

Under a separate heading, he wrote "Intimate Relationship":

Baby Steps
    Step 1—Having fun, people around, enjoy myself—
        "she's fun"
    Step 2—a buddy, first couple of steps; relationships
        develop

In his journal, he also asked himself, "Can I see myself playing hockey, married, and have kids??" He answered: "Up to now, NO." Then he mentioned the names "Babcock, Gélinas, Valiquette," three families that he liked and respected, and asked, "What would Steve married with kids look like?" He set out what he thought it would take for him to make marriage work:

> Defining myself as a family man—good father (trying my
>   best); good husband; caring, supportive.

He also wrote:

> I know there is an incredible woman for me.
> The right one is worth fighting for.

He wrote down the qualities he believed he brought to the Blackhawks—"gritty, hard working, puck moving, physical defence-man, money in the bank in my own zone, shots through on net, tough to play against, positive leader"—and called these collective qualities, and himself, "TIGER INC."

A few days later, Steve's mood had changed:

> trapped—scared
> Why am I doing this? I need to know why I'm doing
>   something.
> Some parts of me that feel I'm undeserving

A few days after that, he wrote:

I need to <u>Decide</u>
Decide to make Chicago MY CITY
Decide that I belong in this uniform, team, with these guys
I belong here, this is my home, this is my city . . .
I picked it, it picked me
ROOTS ROOTS ROOTS . . .
Something's gonna happen to magically make me belong
Know it in my heart that I belong
If I'm gonna be here, BE HERE

The day before the Blackhawks opened the season against Dallas, he wrote:

Lace up tomorrow—own the fuckin' ice—own the
   arena—own the United Center . . .
Skates, grounding to the core of the earth, powerful
   uniform, I am what I decided to be, this is my
   home—claiming it

Under these words, he signed "Steve."

For Steve, it was a new city, new team, new teammates, new friends, new places to go and things to do, new habits, new life.

The year started well. He played the first thirty-eight games of the season, until January 2, when he missed the first of three games due to a back injury. The Blackhawks were 24–10–4 at the time. He was playing about fifteen minutes a game, normal for a 5–6 defenceman—on regular shifts, not on power plays or killing penalties. Duncan Keith, the previous year's Norris Trophy winner as the NHL's best defenceman, was playing around twenty-seven minutes a game; Brent Seabrook, his defence partner, about twenty-five; the team's 3–4 pairing, Nick Leddy twenty-two, and Niklas Hjalmarsson twenty. Steve's defence partner, Sean O'Donnell, was playing less than fourteen.

Steve was playing the way he had always played: competitive, determined, physical; more difficult to play against than he was punishing; more effective than he was dominant. He was hard-working on the ice and off: in practice, in games, in the gym; and in the community, making frequent fundraising, awareness-raising appearances for Chicago-area charities on behalf of the team. He was a good teammate. He still made the occasional gaffe.

He kept a record of almost every game in his journal. On the left-hand page, he visualized the contest ahead, what the team expected/hoped of him and what he expected/hoped of himself. He began his entry for each game the same way:

> I am so happy and grateful now that:
> I had 3 shots on net tonight and it made me feel EXCITED
> I had 3 hard hits tonight and it made me feel POWERFUL
> I blocked 3 shots tonight and it made me feel PRESENT
> And I was +1 tonight and it made me feel HAPPY
> I did all of these from an athletic stance.

A few games later he added another phrase that he would repeat for each game:

> I had a goal and/or an assist tonight and it made me feel
> AWESOME.

Soon after he added another:

TRUE GRIT
This team is Mine, this shift is Mine, the puck is Mine
Defending—this zone is Mine
Attacking—this goal is Mine
I HAVE ARRIVED!!!

Also on the left-side of his journal page, beneath his expectations/hopes for the game, he wrote down some of the good things he had done that day. On October 13, 2011, before the third game of the season, at home against Winnipeg, he noted: "I prayed for my family today and I gave my brother 2 tickets."

On subsequent game days, he noted other things that might seem too small to matter:

I bought food/drinks for a homeless guy yesterday.
I gave Car Bomb some chew.
I left tickets for Oscar the barber.
I held a door for a lady at the rink.
[And, on November 11]
I gave thanks to all veterans past and present and those who
gave their lives in Canada's and USA's military. I also
offered Car Bomb lunch today.

On the right-hand page, after the game, he wrote his observations about how he played that night. He would mention the good things he did and what he needed to do. He offered no self-criticism. For the game against Winnipeg, he noted:

I played well, I was simple with my positioning and the
puck. I was physical. I can join the rush more, use my

speed to aid in the attack. You're a good player, know it.
BE IT.
OWN IT.
My town, my team.

The stats sheet that night showed he had one shot on net, two hits, two penalty minutes, was minus-1, and played fifteen minutes. The Blackhawks won, 4–3.

Steve scored his first goal for the Blackhawks in Columbus in mid-November, the sixteenth game of the season, also recording an assist and a fight—Steve's first and only Gordie Howe hat trick, which gave him half as many as Howe himself had in his entire career. Three nights later, he scored two goals; three nights after that, another goal.

But things were going less well than they seemed.

A year before Steve signed with Chicago, Mike Kitchen had been hired as a Blackhawks assistant coach. A defenceman himself in his playing career, Kitchen was small and had limited offensive skills. He'd known that if he were to make the NHL, and stay there, he would have to work hard, train hard, be disciplined and smart, accept the role of the 5–6 defenceman, and play a simple, uncomplicated game. He played eight NHL seasons.

Kitchen had also been an assistant in Florida when Steve was there, and had worked with Steve to simplify his game. He would remind him when he was trying to do too much, and Steve would shake his head. He knew; and Kitchen knew that he knew. And the next time it happened, Steve would shake his head again, and know again, and Kitchen would keep working with him, and grow more frustrated. It was unimaginable to him that Steve couldn't do what he so clearly needed to do. *Know what you are and what you aren't, for heaven's sake. Do what you can do and don't do what you can't. What you can do is good enough!* Both Kitchen and Steve were good guys. Both were team players. They should have been a good match.

In Chicago, Kitchen again worked with Steve to simplify his game. *This is the Chicago Blackhawks, not the Florida Panthers. The team has Toews, Kane, Keith. . . . When it needs a big play, let them make it. Don't even try.*

Steve had never played on a team like the Blackhawks before. The Flames in 2004 had come out of nowhere, giving the players and fans an unexpected gift, but that had created only hope, not lingering expectations. Chicago had won a Stanley Cup two seasons ago, and the team was getting better. They had the chance to win every year.

It had seemed that same way for Chicago once before, in 1961, when they had the great Glenn Hall in goal; the league's best defenceman, Pierre Pilote; the scoring champion, Stan Mikita; and the game's most charismatic player, big-shooting Bobby Hull. But after winning the Cup that season they didn't win again, and by the mid-1970s, Chicago was beginning to slide. Their owner, Bill Wirtz, who had inherited the team from his father, Arthur Wirtz, didn't like the reality of the WHA, so he ignored it. Later, he didn't like the reality of free agency or of TV broadcasts of local games, so he ignored them. This was *his* team. Nobody was going to tell him what to do. And in winning every battle he fought, Wirtz turned a great hockey city into a sour, deeply disappointed one.

The team became competitive again in the early 1990s, with Chris Chelios, Jeremy Roenick, Steve Larmer, and Ed Belfour, then inconsequential for a decade until Jonathan Toews and Patrick Kane, and coach Joel Quenneville, arrived to join the rapidly improving Duncan Keith. Crucially, Rocky Wirtz, who believed in spending money to make money, took over ownership of the team from his late father. Joy returned to Chicago hockey, and with joy came fans, money, and success—and with them more fans, more money, more success, and fist-pumping pride. The Blackhawks had learned how to win, and had learned what winning feels like. They wanted nothing less, and would accept nothing less. They were hooked, and the fans were, too.

The loss to the Canucks in the first round of the 2011 playoffs had shocked all of them. For the Blackhawks, the next season couldn't begin soon enough. It was as if ownership, management, coaches, and players were on a quest. Good guys on a good team are forgiven for bad moments. Good guys on a Cup-questing team are not.

At Steve's first lapse, Kitchen's memories of Florida flooded back. He didn't trust Steve, and Steve knew it. "Value me and I feel valued," Steve had written in his journal.

Steve's journal entries got longer. He was in contact with Gisele more frequently. His lapses didn't stop.

CHAPTER EIGHTEEN

Steve was every girl's dream, until he wasn't. He was smart, funny, good-looking. He was strong and gentle; loud and quiet. He had a body that turned heads. He was cool, because he just was. He didn't need to be with a crowd, at the centre of every moment, to be the most important person in the room. But often, it would turn out that way. He could be alone, reading, hour after hour on his laptop, at the cottage he had on a lake in the middle of nowhere. He liked to be alone. But he also liked to be with others. With his teammates and buddies, and with girls. With lots of girls—and at different times in his life, with one girl. With Casey, Helga, and Star, and others. If Steve saw you as special, you were special. From the very first moment.

He always jumped in with both feet. He couldn't do enough. He stared into the eyes of that special girl, he took in everything that was in her. He wanted to know everything about her, what nobody else knew, what she had never told anyone, what she might not have even known about herself. And he listened, to every word she said.

He saw himself as spiritual. He believed in religion, but he thought there were so many other things to believe in, too. In every experience, in every person, there was something to learn, and in every next experience and in every next person. And in himself, too, and every tomorrow. He wasn't a seeker, looking relentlessly, endlessly, for the meaning of life. He was an explorer. He wanted to go everywhere, do everything, meet everybody. He wanted to learn, and find meaning, but not *the* meaning. His quest wasn't out of emptiness, but delight. Fascination.

And at some moment he realized he had the right life to do all this. He had money. He had time. He would have to pick his spots; he had games and practices and hours in the gym to attend to. But he had time in between, and between seasons, at Olympic and All-Star breaks. From a team's last game to its first practice back, that might only give him three days, but that was seventy-two hours. He could charter a plane; he could go anywhere. After all, he didn't need much sleep, not when there was something to do. And besides money and time, he had access to this life because of *him*. Not because he was an NHL player, because most people didn't know his name or what he did, and he didn't tell them. It was because he could talk to anyone, because anyone wanted to talk to *him*. Because he was nice, and friendly, and smart, and because he wanted to know about them, because he made them and what they were doing more important than himself.

But to be the special woman in Steve's life wasn't easy. High reward; high risk. If he was all in, she had to be all in. If he was to get so close to her, to know so much, to inhabit her, she needed to inhabit him. And for days, sometimes weeks, or even months, that's how it might be. His father had believed that you learn everything you can in a job within two years, and then it's time to move on. Steve had a need to move on, too. But in his mind, he didn't. He was a romantic. A believer. He wanted that special girl. He wanted a family; he loved kids. But then within a few days, or a few weeks, or a few months, he'd find something that wasn't quite right.

Keats recalls this one girl, she was "perfect," Steve told him. Steve was so excited. He couldn't do enough for her. "Then he said to me, 'But she's got this toenail. I can't deal with that. She's got a bad toenail.' And I say to him, 'Monty, what's wrong with her toenail?' and we start talking about a toenail. 'Is it her big toe, or her small toe?'" Big reasons or small reasons—for Steve, there were always reasons.

He might see one of his former girlfriends a few years later; he might even think about how he had really blown it with her. He never believed there wasn't a special girl out there for him, or that he wouldn't find her.

That season in Chicago, Missy Holas moved in with Steve. The two of them had met at a party in 2004 in Calgary during the Flames' Cup run. At the time, Steve was a college-age kid living an adult life with adult responsibilities every night the Flames played; and living a college-kid life of parties, alcohol, and drugs every night in between. At least, years later, that's how Missy remembered him, and she hadn't been impressed. Missy was a chiropractor and was living in Atlanta. She had clients on several NHL teams that she would see when their teams were in the city to play the Thrashers. She was often in the dressing room area, and over the years, when Florida, Anaheim, or Buffalo were in town, she and Steve might walk by each other, smile, and say hi.

Then Missy moved to Chicago, and in 2011, so did Steve. Steve's body, like that of every player, would begin a season fit and healthy, and under the weight of accumulated spasms, tightnesses, and other insults, gradually break down. He was looking for a chiropractor, and had heard about Missy from other players. He became one of her clients. She manipulated his muscles and joints, and gave him acupuncture; he asked her questions. "He wanted to understand everything," Missy says. "Why his body was working, or why it wasn't." She began preparing for his appointments. What was he going to ask? He also wanted to know everything about her. Missy was used to pick-up questions; with Steve, it was like he really wanted to know. She found him fascinating, and exhausting.

Missy liked athletes. She liked their energy and purpose, their physicality, and they liked her. She was cute and fun. She talked like a jock and thought like a jock. She liked to hang around athletes and they liked to hang around her. She was not like any chiropractor any of them had ever met. She was one of them.

Steve and Missy were not girlfriend and boyfriend. Steve saw many other women; Missy saw other men. They were soulmates, kindred spirits, and roommates. They could spend hour after hour with each other, doing exciting things, or nothing at all. It's as if they hung out in each other's minds. And unlike almost every other woman Steve had met, Missy could keep up with him. He could do ten things a day; she could do ten things a day. He could find every next thing totally interesting; so could she. And while Steve could talk for hours and hours, Missy could talk for hours and hours *at warp speed*. When it came to talking, she was Rogers to his Astaire. She could do everything he did, but going backwards and in high heels. At times, it was too much for Steve. "Quiet time, Missy," he would say to her. Many of Steve's friends thought she was perfect for him.

Missy had met Steve first during his drinking years in Calgary, saw him infrequently during his healthy years, and got to know him during his increasingly fragile years. He was still good that first fall in Chicago, Missy recalls.

Then the concussions came.

On January 8, the Blackhawks played Detroit. When the game was over, Steve went to see Dr. Michael Terry, one of the Blackhawks' physicians. He had been hit in the face with a punch, he told Terry, and as Terry later recorded in his notes, had "a brief period where his consciousness was altered," where he "felt like he was a little hypoglycemic" and "a little bit hazy." Terry put him through some tests and wrote:

> [H]e was able to perform serial 7s. Three word recall was intact. His modification of SCAT was otherwise normal although he was unable to do 3 number reverse recall and 4 number reverse recall without prompting. Otherwise normal neuro exam. Cranial nerves 2 through 12 are intact. His balance is normal. His affect was normal.

In his notes, Terry concluded: "A/P [Assessment and Plan]: concussion. We are going to put him through our protocol for return to play." On Steve's Fitness to Play Determination Form, which he signed, Terry wrote: "Disabled."

Two days later, January 10, before the Blackhawks game against Columbus, Steve was given a neuropsychological examination by Dr. Elizabeth Pieroth under the NHL Concussion Program. Steve described his injury and symptoms to Pieroth, as he had done with Terry. Pieroth later wrote in her report that: "The player was a poor historian regarding his concussive history. He stated his most recent concussion occurred February 2011 but he could not recall how he was injured. He denied loss of consciousness or retrograde/ anterograde amnesia and experienced only headache and neck pain. Steve believes he missed 2–3 games and thinks this was due to neck pain." Further, Pieroth reported, "In 2009 he was hit in a practice and missed 2 games but wasn't sure if he had actually suffered a concussion at that time. The player also reported that in September 2000 he received a blow to the head and was out of play for one week but does not recall what symptoms he experienced or how long [they] lasted." Pieroth checked Steve's recollections against those he had reported on his earlier NHL tests and found them "not entirely consistent." In addition, "Steve cannot state if these were all concussive injuries. He denied any other head injuries or significant medical history."

Pieroth went on, describing his present symptoms after the hit two days before. "Currently Steve did report trouble falling asleep and irritability but stated this was secondary to personal issues and not related to his concussions. He has also completed the Blackhawks' exertional protocol without eliciting any symptoms. Therefore he can be cleared to play in this evening's game with the team physician's approval."

Pieroth concluded her report:

Steve was also provided with education on concussive injuries and we discussed the current signs of multiple concussions. I explained to him that it is difficult to determine if he has demonstrated increased recovery time or increased vulnerability to concussive injuries given his poor recollection of his past injuries. However, given his report, the player does seem that he has recovered along expected lines from his previous concussions. Steve stated that he understood my concerns but is comfortable with assuming the risk associated with continued play at the professional level.

Steve's Fitness to Play Determination Form, again signed by Dr. Terry, reads: "Not disabled." Steve played that night against Columbus.

Almost a month later, on February 8, Adam Jahns of the *Chicago Sun-Times* reported that Steve had left the game the night before against Colorado in the second period, "and didn't return because of an upper-body injury. He underwent X-rays for his injury before the Hawks left the Pepsi Center. His injury isn't thought to be serious. 'He's doing OK today,' [Blackhawks coach Joel] Quenneville said. 'We'll see how he is [tomorrow].'"

The next day, Steve was put on the injured reserve list.

Nearly three weeks later, having seen Dr. Terry again and still feeling symptoms, Steve decided to seek another opinion. He went to see Dr. Jeffrey Kutcher at the Michigan NeuroSport Clinic at the University of Michigan hospital in Ann Arbor.

Kutcher, in his report, filled in some of the details between Steve's initial injury on January 8, his examinations by Terry and Pieroth, and his injury against Colorado on February 7. Kutcher noted that after his January 8 injury, Steve played the next four games "without any symptoms or difficulties." Then, he wrote:

On January 18th, he was struck in the head by an
opponent's shoulder which caused a more significant
constellation of symptoms. For the next 2 weeks, he
felt forgetful, emotional, and 'out of it.' He was beginning
to sleep poorly. He was having mild diffuse generalized

head pain. Despite this, he continued to participate in hockey.

On February 3rd, Steve was in a game when he was hit from behind causing a whiplash-type of movement. This resulted in perhaps a short duration of loss of consciousness. He continued to play, however; and on February 7th, while playing a game at Colorado he was involved in a more subtle hit, but one that caused an immediate flash of a green fence in his vision. He played the rest of the period, but then removed himself from participation.

A "green fence" is one of a variety of visual images, which appear in different colours, that concussed people recall seeing at the time of impact.

Steve had been injured first on January 8, then again ten days later on January 18, then two weeks after that on February 3, then four days later on February 7. His meeting with Kutcher occurred twenty days after his last injury in Colorado. Kutcher described his symptoms that day:

> "Currently he is continuing to experience focal pounding headaches that last anywhere from 5 min. to 2 hours. They occur sporadically, but also with minimal exertion. Symptoms at baseline have improved other- wise, but he still describes having problems with sleep, mood, and appetite. As he has improved, he has attempted to return to physical activity with two trials on a stationary bike. One was on February 23rd for 10 min. and when his heart rate got to approximately

120 he had a significant increase in head pain. He again
tried this on the 24th with essentially the same results."

Kutcher also gave Steve a general examination, and described what
he found as "unremarkable." He put him through additional tests
while he was at rest, then after twenty-two minutes of exertion on
a stationary bike. The next day he gave him the same bike test,
then some agility drills in the gym followed by drills on the ice.
"He did very well tolerating exertional levels much higher than
previously noted," and "without any significant increase in symptoms," Kutcher wrote.

He concluded: "I was encouraged today by his performance
and the rather subtle symptoms that he expressed." Kutcher added
a note of caution: "At this point, while I'm encouraged, I would like
to be very careful moving forward," and he suggested "we progress
along a very careful rehabilitation program that stresses both
increased exertional levels as well as agility, movement, visuo-
spatial tasks, and the cognitive aspects of playing hockey."

Kutcher added: "[W]e discussed the possibility of medications
to help. He would like to forego any medications at this time, but
we will continue to monitor his symptoms and he may reconsider
this in the future."

Steve returned to Chicago. On February 29, he was examined
by Dr. Terry. The doctor's dictation note was less encouraging:
"[Steve] still is feeling foggy. He has not noticed a good deal of
change but he has been exercising a bit. He says that he stops
when he is symptomatic." Nine days later, Steve saw Terry again.
The doctor's dictation note reads: "[Steve] says that he has been
feeling better. Will have an occasional feeling of vertigo or dizziness. An occasional headache. He said that they are both very rare.

Overall he said he otherwise feels essentially normal with no focal symptoms."

After Steve's initial injury on January 8, his journal entries became more sketchy and sporadic. On March 14, he wrote:

1. Gratitude—health
2. Gratitude—game
    Remaining Empowered
    Concussion—stop the noise

Later the same day, he asked:

What is this teaching me about being a competitor?
- keeping things simple
- how to handle reality
- accepting fallibilities . . .
- overcoming adversity

More often, he began a thought in his journal and didn't complete it. He wrote of his gratitude for the chance to play with "this unique club": "I'm good, it's fun, there's tough patches but that's OK. [My] capacity for survival is phenomenal. Pat self on back, it's OK. I bring a lot to the team. I know my being there boosts that environment."

A week later, on March 21, Steve saw Dr. Terry again and told him he was feeling better. Terry noticed no symptoms, and noted that Steve's modified Standardized Concussion Assessment Tool (SCAT) was "normal." Steve's Fitness to Play Determination Form, signed by Terry, read: "Not disabled." Four days later, on his dictation note, Terry affirms Steve's status: "He has been treated for a concussion. He presents today with no symptoms. He completed his exercise protocol and impact testing and passed both. He is symptom-free and doing well. He is therefore clear to return to play."

On March 27, 2012, thirty-five games after his initial injury, twenty-three games after his fourth injury against Colorado, Steve dressed against New Jersey. He played four minutes and twenty seconds. In the third period, playing on the wing, he crashed the Devils' net and got an "inadvertent elbow" to the head from defenceman Mark Fayne.

The next day, Steve's Fitness to Play Determination Form, signed by Dr. Terry, read: "Disabled."

Steve never played in the NHL again.

It never seemed it would happen this way. Steve's return to play was always a matter of time—time to allow things to settle, time to let the brain heal. A different doctor, a new treatment, a different understanding and approach to concussions, and to paraphrase what Steve had written in his journal, "Something's gonna happen to magically make me *better.*" He had always gotten better before. Athletes get better. You feel, you deal, it passes, you get on with it. That's how it had been with his back, his knee, his neck. That's how it had been with his head. Headaches, dizziness, fatigue, sensitivity to light—symptoms that felt like they were going to last forever always went away. No scars left behind, nothing he could see in the mirror, nothing anyone else could see even on MRIs, no indications that anything had happened. It was only when he was asked by doctors about his medical history that he even remembered all those other hits to his head, that they might have been *something.* The time in minor peewee, that other time in junior. Those times in Calgary when he was trying to make the team, when he did a face plant on the ice and cut up his nose and cheek, when he got knocked silly by an elbow, when he got sucker-punched in a bar. The time in Florida when he ran into an opponent's helmet with his face and broke his nose again. Then, in Buffalo, another elbow, a stick, more cuts, another break to his nose. In every instance the injury he *thought* he received was a break or a cut, not dizziness or headaches. It was the same earlier in the season when he slid into the boards with his face and fractured his zygomatic arch and temporarily lost his hearing. He'd had a brief loss of consciousness; but isn't that just what happens when you break your cheekbone? If he had a banged-up shoulder *and* a headache, which injury was he going to focus on? Players get hurt game by game; all his big head-hits, until that season, had happened months and years apart. He'd gotten better in between. He was always fine. He had a game to play.

Athletes have their own kind of relationship with pain. They play because they are so absorbed in playing that they don't notice injuries when they happen. The soldier who is shot keeps on going because the imperative to go on is so much more important than the imperative to fall. The explanation for such a miraculous act is purpose more than courage.

So players play. And players expect other players to play. Someone goes down in a hockey game and is helped off the ice. "He'll be back," the announcer says. "He's a hockey player." And when a hockey player does come back, most often he is fine. Time heals. As a player, you learn quickly: Where do you want to spend your healing time? At home, moping around, feeling the pain, having nothing to do to distract you from it? Or on the ice, with your buddies, who admire you just a little more because you are there, doing what you love and doing it for the team? Like Keith Primeau and Marc Savard did. Like Steve always did.

Steve played for a month between January 8 and February 7 with concussion symptoms that kept recurring. Four times. Four separate hits. January 8, January 18, February 3, February 8. Why? Because he didn't feel *that* bad. Because he was able to persuade himself that he didn't feel *that* bad. Because he had played other times with a wonky shoulder, so why not a wonky head? Because he was tough and everyone knew he was tough, and he liked that he was tough. Because as lousy as he felt during those other twenty-one and a half hours of a game day, for the two and a half hours he played, he didn't feel lousy. He felt *the game.* And for those games between January 21 and February 7 he especially played because the Blackhawks had lost five in a row and his teammates needed him. Because he needed to prove something to them, to Quenneville and Kitchen, to himself, to TIGER INC., to Stan Bowman, the Blackhawks general manager, who had signed him. Because not to play was unthinkable.

When he was a kid, he didn't know anything about this long-term injury stuff, or even think about it. Why would he? He just played. Now he knew there might be some effects, but he didn't really know—*he had no idea*—what those effects might be, no matter how much Dr. Terry and Dr. Pieroth and all the other experts told him and what he said back to them. *I know,* he'd say; *I understand.* But he didn't. How could he? He'd had no idea he would feel like this. None. He'd had no idea this might go on for weeks. For months. He'd had no idea, he still had no idea. This might go on and on and never stop. This might be his life. This might be him.

When he was younger, he knew: *I have to do everything in my power to make this life happen.* Now at thirty-two, he knew: *I have it, and I must not, I will not give it up.*

The updates in the media about Steve's injury became less frequent. The Blackhawks were in a playoff fight; Steve would be news only if he returned to the lineup, and no one believed that would happen. The team finished sixth in the Western Conference. Two weeks later, their season was over; they lost in the first round to the Coyotes in six games. For a team with such possibilities and appetites, this had been a bad year. Steve had been signed as a tweak to a lineup that needed only a few tweaks to return to Stanley Cup glory. His injuries, and the new needs of the team, now made his future with the Blackhawks very uncertain.

Two days after Chicago's season ended, on April 25, Steve saw Dr. Terry again. Later, the doctor dictated this note: "[Steve] remained symptomatic from his concussion and we will continue to monitor him and take him through our protocol when it is appropriate." Again, Steve was listed as: "Disabled."

Four days later, he travelled to Ann Arbor to see Dr. Kutcher, who put him through two more days of tests. After the second day, Kutcher wrote: "He is doing well today with no significant symptoms. . . . [His] examination again today was normal including intact mental status, cranial nerves which were intact to specific testing of each cranial nerve. Excellent balance and coordination was also noted again." Kutcher said that he was encouraged by Steve's improvement, the brain MRI hadn't shown any obvious signs of trauma, and he was cautiously optimistic about Steve's return to hockey.

Early in June, Steve went to see Dr. Ted Carrick at the Carrick Institute Life University Neurology Clinic in Marietta, Georgia, near Atlanta. Carrick is a chiropractor specializing in what he calls "chiropractic neurology." He had treated Sidney Crosby and been featured prominently in the concussion press conference in

September 2011, in which Crosby's progress was outlined. Two months later, Crosby had returned to play, then was sidelined eight games later after a minor hit, then played again on March 15, twelve days before Steve made his brief return against the Devils. Carrick had been recommended to Steve by Andy O'Brien.

Steve related his concussion history to Dr. Carrick as he had done to Terry, Pieroth, and Kutcher, with only a few added details. He told Carrick that after his initial injury in January he became "forgetful, emotional, and . . . small things would bother [me]," and that when he continued to play during the period of his three successive injuries, he experienced dizziness and problems with his balance and reflexes. Now, in June, Steve described himself as "lightheaded, with a slightly drunk feeling, and a feeling of disconnection." He told Carrick that by mid-afternoon every day he felt "crushed for the rest the day." He was taking no medications, but had been "more emotional and depressed."

In his report, Carrick also listed a catalogue of Steve's injuries, including "chronic problems with the left eye," a "leaky gut," a "herniated disk" he had suffered in 2007 and aggravated the previous year, and a broken "zygomatic arch," as a result of which, seven months later, he still had a "hard time opening his mouth." After his examination, Carrick strapped Steve into a gyroscope, where he was spun around repeatedly for "whole body rotational stimulation." Later, Carrick put Steve on the ice for movements that included "rotations, stops and starts with alternating head and eye movements."

Carrick wrote in his report: "I've attended him for 4 days while in rehabilitation. . . . He has made significant improvement in his neurological function. He has some minor observable deficits but performs above the range of normal subjects. He should be able to play hockey."

When Steve left the clinic, he called Vally, even more excited than usual. He tried to describe to Vally his experience with Carrick, and the gyroscope. "Vally, Vally, it's an enclosed . . . have you ever seen one of those hamster cages where they have to run around on that wheel? It was a chair, and you were strapped in, and it gyrates. It moves you up and down and around." Vally asked him how much each session cost. Steve told him $1,500. Vally was stunned. "But, Vally, Vally, it cures you, it cures you," said Steve. "We've got to buy these machines. We've got to set up our own clinic."

Less than two weeks later, the *Chicago Sun-Times* reported that Stan Bowman, wanting to give the Blackhawks some cap room, was "looking for takers" for Steve. Four days after that, the *Chicago Tribune* wrote that Steve would be ready for training camp.

"I feel like a whole new man," Steve is quoted as saying. "I'm as excited to get this season started as any other, if not more so just because I feel like I'm clear and healthy more than I've been in years. That's really exciting for me."

On July 20, less than a month later, Steve was again examined by Dr. Terry. Terry's report read: "He has been doing well. He has been exerting himself and continues to work out. He is building his activity level at a reasonable pace. He does still report some symptoms including dizziness, fogginess, haziness, and these are generally associated with workouts but not consistently so." Terry writes in summary: "We will put him through the exertional pro- tocol . . . [and] while he is still having symptoms we will hold off on clearing him for play. We will continue to follow him along."

Meanwhile, the CBA that had been reached after the lockout was about to expire in September, and talks between the league and the players had become more regular. Eight years earlier, the inabil- ity of the NHL and the NHLPA to reach an agreement had resulted in a lockout, a lost season, the implosion of the NHLPA, and the league getting its salary cap. Three days after his appointment with Dr. Terry in July, Steve spoke to several journalists at the NHLPA Golf Classic just outside Toronto, about the state of the CBA negotiations, setting out clearly, carefully, where things stood. He was mindful to compliment both sides on how things were proceeding, and offered few specifics, because few existed; yet he didn't sound evasive. He looked at each questioner, and paused before he spoke each time. He looked patient and respectful; healthy and strong.

It was still the summer. California, Toronto, New York—he could go wherever he wanted to go. It was good times without bad times. Some of his concussion symptoms came back, but then they disappeared. That's what was important: he had been cured. And if they did come back again, he could be cured again. Sidney Crosby had been out for a year and had played in the playoffs. Crosby had gone to Dr. Carrick and had done the gyroscope. Steve had, too.

And there was lots of time. No games were being played, or missed. He wasn't letting anybody down because of his injury. He had no rush to get back. He was training as much as anybody, and more than most. He was getting tested, being treated, and this time the impending lockout would be on his side. The longer the lockout went on, the better. And the best news? No setbacks. He was fine unless—and until—he wasn't fine. Things weren't where they needed to be, but they were where they had to be at that moment.

In early August, Steve went to see Vally at his new house in Connecticut. "Oh my god, Monty was Monty," Vally recalls. "He was healthy all the way. Working out. Great shape. Vibrant. Alcohol? No chance. He wasn't drinking. We were totally having a healthy time together." Late one morning, Steve and Vally were sitting out in the sun. "'What do you want to do today?' I ask him. Silence.

Then I say, 'We should get some paddleboards. There's a place in Westport.'" But Westport was thirty minutes away and, as Vally puts it, "We don't feel like moving." A couple of hours later, they still hadn't moved. "Then some guy pulls up to the house, and he's got two paddleboards!" Vally still can't quite believe his own story. "Monty had snuck away, called the store, and had them delivered and paid for. I'm talking about three grand's worth of paddleboards! I said, like, 'Monty, you've got to be kidding.' He's like, 'Hey chum, it's my housewarming gift.'"

And now that they had them, they had to use them. "Monty and I found this really wicked spot on the Housatonic River," Vally recalls, "where it opens up to Long Island Sound. There's a small sliver of a channel; it was like our own private bayou. So we cut through it, and we're in there for a few minutes, and when you get to the end you've got a decision to make. You can go back out the same way, or you can portage over this bird sanctuary. So we decide to go through the bird sanctuary; and on the other side the Long Island Sound picks up again. The whole route is about fifty minutes down and back, and the tide has to be in to do it. If it's out, you hit the rocks. And the portage, nobody else does it. It's just something Monty and I created."

When the two of them got back, they sat around in the sun some more, threw a football, then went back to the beach. "We had a really fun week together," Vally remembers, "and he was as healthy as you can believe."

"Monty was very upbeat that summer," Missy recalls. "Goofy. His brain was good."

Some players, imagining another lost lockout season, were making alternative plans to play in Europe. Steve decided to stay put. He was going to train like the season would start tomorrow, and get involved in the CBA negotiations. At 11:59 p.m. on September 15, the NHL locked out the players. Less than two days later, playing in a pickup hockey game, Steve got an elbow to the head.

CHAPTER NINETEEN

On November 1, 2012, Steve again went to see Dr. Elizabeth Pieroth. He told her that in mid-September he had been elbowed in the head. The hit had "buzzed" him right away, he said, and then, very briefly, he had seen a "green fence," but he didn't have any other immediate symptoms. Later that same day, however, he had developed an intermittent headache, feelings of nausea, and "spaciness, feeling drunk." He took ten days off from his workouts and began to feel better, Pieroth wrote in her NHL concussion evaluation report, and he received neck massages that helped reduce the headaches. He began exercising again, but cautiously, and was able to ride a stationary bike and lift weights without any symptoms. But once he went back on the ice, the symptoms returned.

Three weeks after his injury, he began to engage in light contact in his on-ice workouts, he related, describing it as simple physical "resistance" with the other players. Over the next few weeks, he had increased the level of contact. Pieroth noted that Steve "has completed the Blackhawks' return to play physical exertion protocol and many on-ice practices without any returns of his symptoms [and] last had any symptoms 2–3 weeks ago," which, Steve told her, "were only momentary." Pieroth reported, "[He said,] 'I feel like I can take hits without concern.' He said he was '100% okay with returning to play' and said he had 'no reservations about playing this game.'"

Pieroth noted, too, that Steve had completed the NHL concussion protocol, that his scores were consistent with his baseline results with only minor exceptions, and that there was "no other evidence of significant cognitive impairment secondary to his injury. After reviewing Steve's history," Pieroth continued, "his description of this most recent event, his recovery over the past 6 weeks and his current test results, I believe he can be cleared to return to full-contact play without restrictions. This is, of course, with the approval of the team's head physician."

According to Pieroth, she and Steve also "spoke at length about the current signs of multiple concussions," just as they had done when she first examined him ten months earlier. She wrote:

> I explained very clearly that there is no way of predicting if an athlete will suffer long-term problems from their history of concussions and repetitive contact to the head during their career. I also expressed my concern about his history and we reviewed the indicators when an athlete has suffered too many concussive injuries to continue playing in a contact sport. Steve expressed good understanding of this information and stated he does not plan to consider retirement from the sport due to his concussions at this time.

On the same day, Steve saw Dr. Terry. He told Terry that he had been working out without any difficulties and was back to feeling normal on the ice. He said he had taken an impact test and had no concussive symptoms, and had seen a vestibular therapist for an evaluation. According to his report, Terry found that Steve had no concussive-related issues remaining. Terry also reported that he discussed with Steve "long-term plans, long-term risks of continued play with recurrent repeated and prolonged concussions and his future safety. He understands these very well; he is very knowledgeable about the subject and has been for some time. He expressed a willingness to be cleared to return to full contact and has completed our protocol completely, and therefore we will allow him to return to play with no restrictions at this point."

"Not disabled," Dr. Terry wrote on Steve's Fitness to Play Determination Form.

Four days later, Steve was bumped again.

Dr. Pieroth described the month that followed in a report she wrote in February 2013:

> [O]n 11/5/12 [Steve] received a mild "bump" (described
> as being "grazed": on the left side of his head) during a
> practice and developed symptoms of mild headache,
> mental fogginess/spaciness and neck pain. He believed
> the headache was related to a cervical/muscle issue and
> did feel better with treatment of his neck. Steve flew to
> Calgary on 11/13/12 for therapy on his neck, which he said
> lessened his symptoms overall. He skated on the 19th
> and 21st without any issues. However, on 11/21/12 he
> was elbowed to the face, which caused a return of his
> symptoms (mild).

The player flew to Phoenix on 11/25/12 to skate with other NHL players during the lockout. He received chiropractic treatment and felt better. He was working out/skating daily but again took a mild bump to the head (11/27/12) and he felt "spacey" again. Steve took a few days off and returned to skating on 11/29/12. On that date he collided with another player [both of whom were skating backwards], which caused a return of symptoms."

November 5, November 21, November 27, November 29.

Steve returned to Toronto for Christmas and stayed with his brother Chris. For two days, Chris recalls, he just sat on the downstairs couch. "He couldn't come upstairs. He was bothered by light. Usually, he'd chirp you, and when I say chirp, I mean bust your balls. He didn't even want to talk. He was just kind of hunched over. I was saying to him, like, 'Just lie down, go for a walk, get some fresh air.'"

On December 28, he flew back to Chicago and was examined by Dr. Julian Bailes of the NorthShore Neurological Institute. "According to Dr. Bailes' report," Pieroth wrote later, "[Steve's] neurological examination was normal. Dr. Bailes recommended antidepressant medication, which Steve refused at that time, and various nutritional supplements. Dr. Bailes concluded, 'I believe that based upon his present condition, Mr. Montador could attempt to return to play using a customized and slowly progressive protocol unless he sustains recurrent symptoms. He has a normal neurophysiologic and neurologic examination.'"

One day later, Steve flew to California and checked himself into rehab at The Canyon Rehabilitation Center in Malibu. He was

accompanied by Dan Cronin, director of counseling for the NHL/
NHLPA's Substance Abuse and Behavioural Health Program.
Steve's admission form read:

> Patient had been 7 years sober and used marijuana one
> time a year and a half ago. He is currently suffering from
> increased depression he believes to be from head traumas
> from his line of work. In the last month and a half specifi-
> cally the symptoms have gotten worse. Patient is taking
> supplements to help him with this and will be with us for
> a limited amount of time. He may need to leave within a
> week to two weeks for work reasons. [The "work reasons"
> relate to the possible end of the lockout and the resump-
> tion of the 2012–13 NHL season.] Has had some suicidal
> thoughts and has had cravings for cocaine. Unhealthy
> sleeping patterns and lack of motivation. Patient also has
> childhood issues of not feeling good enough. His profes-
> sion has given him a passion in his life. Has not had very
> long relationships or successful ones. . . . Lack of feelings.

Steve was asked a series of questions as part of his admission,
to which he offered his answers:

> Can you stop drug use without a struggle, no cravings or
> urges? *Yes*
> In the past could you stop drug use without a struggle,
> no cravings or urges? *No*
> Have you ever thought of killing yourself? *Yes*
> Have you ever attempted suicide? *No*
> Have you had thoughts about killing yourself today? *Yes*

Are you thinking about killing yourself now? *No*
To what degree are you [feeling] hopeless (out of 10)? *7*

The next day, Steve wrote in his journal:

I'm frustrated, sad, and scared. I guess I do feel.
I'm not sure I don't want to use, drink, or die. But they
all seem like great options.
Stuck, will I ever get physically better?

Two days later, on New Year's Day, he wrote:

Things I "should" be grateful for
• life, family
• friends
• some form of health
• some form of cognitive ability
• ability to provide for self/family
• sobriety

And then on the next day, this:

Things I am and/or should be grateful for:
• feeling better this aft, why? Don't know, something
lifted.

His next journal entry is about the CBA negotiations. With almost
half the season gone, the NHL and NHLPA were reaching a critical
stage. Later that day, Steve was on a conference call with the other
players. No one knew where he was, or what condition he was in.

The rehab centre's forms offer observations made by Steve and by staff members. From Steve: "Sick and tired of feeling this way. Unsure if it will change." From staff members (about Steve): "Just tired metaphorically and literally."

On January 4, Steve again wrote in his journal about the CBA negotiations. His notes were cryptic, and several issues remained to be resolved, but the sides, it seemed, were closer. On January 5 and 6, he wrote down the main provisions for the prospective deal. While his entries on most other subjects tended to take up one page or less, his CBA notes covered ten pages. Buried in them were three short lines:

Transition—compliance buyouts,
2 compliance during 13/14—14/15
2 total—100% off cap

Under the proposed new CBA, NHL teams would be given two years to buy out the contracts of two players whom they no longer wanted and couldn't trade. The players would be paid the full amount of their contracts, but the contracts wouldn't apply under the team's salary cap, allowing it to improve itself in other ways. Steve's head injuries had made it likely his NHL career was over. Unbeknownst to him, these three lines made it certain.

As his treatment required, Steve wrote a goodbye letter to substance abuse and destructive behaviour. He wrote it on the same days that he jotted down his CBA notes:

To all the substance abuse and destructive behavior in my life, beat it!
    To all the substance abuse and destructive behavior in my life, where would I be without you???

I owe you so much, thank you for bringing me to
Hell, thank you for making me an alcoholic/drug addict,
for helping me hate myself, never accept myself, and never
connect with another human being on a level I can feel.
Thank you for teaching me how not to treat a lady, a
parent, a friend or any being. Thank you for locking me
inside my own mind, locking me out of my heart, and for
casting a shadow on my days; for that anxiety, hopeless-
ness and fear. I am especially grateful you have hurt me,
but not broken me. For that, I'm definitely grateful.

For if there is light, I see it now. Sun has shone and
chains are broken and I can walk freely within the
moment with acceptance never dreamt of, with joy filled
with laughter, within my own skin. To love you is to
appreciate you, you showed me the way. It was the wrong
way, but a way I know well, one to avoid, with all the love
in the world.

To all of my days in bed, despair in hand, suicidal
thoughts on my brow, you are cast away with a hug. Your
stories will never be forgotten, if not at least appreciated
for their candor. Each moment that passes proves hope is
real, love can be given, and dark days and shadows may
stop by, but leave again.

I resign from your grip.

Your former servant,

Steve

The centre's final report concluded that Steve had made progress at The Canyon, but that his prognosis was "guarded." He was
prescribed a grocery list of antidepressants:

Lunesta, 3mg. one tablet

Seroquel 100mg. one tablet

Seroquel 25mg. one tablet

Zoloft 50mg. one and one half tablets

And, as needed:

Gabapentin 600mg. one half to one tablet

Propranolol 20mg. one tablet (every four hours)

When he checked out, Steve collected the items he had brought with him. Among them were five books: *Lullaby* by Chuck Palahniuk, *The Practicing Mind, Empire of the Summer Moon, Slaughterhouse-Five,* and *Zen and the Art of Motorcycle Maintenance.* Steve also had in his wallet, besides the usual credit cards of someone his age and wealth, a cheque for $20 million he had made out to himself.

On January 6, 2013, four days before Steve left rehab, the new CBA was agreed to by the NHL and NHLPA. The league ratified the deal on January 9; the players on January 12. The same day, back in Chicago, Steve underwent his physical with the rest of his Blackhawks teammates, and failed.

Ten days later, on January 22, Steve went to see Dr. Terry, who later wrote:

[Steve] says he is essentially asymptomatic when working out and with his activities of daily living. He says he does on occasion get a little bit overstimulated and will have some symptoms. He calls these minor and transient. If he shuts down for a bit they go away. The main thing that does this is when he is out shopping. He still remains on his antidepressants. He is still following up externally with psych. He has been working out and has mentioned he has had no symptoms with working out.

Vally had retired as a player the summer before, and was now the goalie coach of the Bridgeport Sound Tigers, the AHL affiliate of the New York Islanders. Steve had called him often since December, and to Vally he sounded different. "Monty was always so excited when we talked, and loud, and full of laughter, and now there was none of this," he recalls. "He was repeating himself. He was talking really slow and low. He was saying things like, 'I'm getting some really tough opinions, and I don't know if I am ever going to feel normal again.' I said to my wife, Chrissy, 'He's not getting better.' I felt that if I didn't get on a plane and help this guy out right away, he's not coming out of this."

Late in January, Vally flew to Chicago to see Steve, who picked him up at the airport. For the entire thirty-minute drive to Steve's apartment, "He was talking and talking about his brain, and telling me all this stuff about what it does, and how synapses work, and all the time he's taking these really big breaths. Monty always took a breath before he'd speak, that was something I thought was funny about him. But this was like *ehhh whoooo, ehhh whoooo*. One after another. It was really distracting. He thought it would get more oxygen to his brain." When they got to his place, Vally says, "I was shocked. In his kitchen, he had this entire counter full, and I mean full, four feet by two feet, of different supplements, all from the vitamin shop, organic *this* and *that*. Oh yeah, and two jars of cashews. And Monty knows what they all are, and what time of day to take them." Vally couldn't believe the state Steve was in. "Then suddenly, he's Monty again. He's telling me, 'Vally, Vally, you've got to promise me one thing. You have to take your omega-threes every day. You need them for your brain. Just promise me.'"

They went downstairs. "He says, 'This is where we're going to hang out.' And that's what we did. He put on some light-sensitivity glasses and we just hunkered down. We were not leaving that basement. It was really dark down there. No natural light. He was on the phone all day, every day, talking to doctors. We watched the Blackhawks game one night. Another, we did go to a restaurant, but otherwise, we just ordered in. We didn't leave. I think he was at his deepest, darkest moment at that point. He was so fragile. He was a different guy."

Steve had always wanted to know everything. Every new thing he learned offered him the promise of something better. Now, the more he knew, the worse things seemed to be. Every new doctor asked him the same questions, gave him the same tests, told him the same thing. The tests said, *you're fine, above-average this, superior that*; but it was the headaches and "spaciness"—what the tests didn't reveal. They were the problem. When they went away, he'd be fine again, he knew that; then he could play again, he knew that; unless they came back again.

He was doing all the right things. Every day he lived smartly: he ate the right food, he got the rest he needed. On the ice, he was so incredibly careful about what he did, and the other guys were so incredibly careful around him. Then he'd get the *tiniest* bump—one that for all of his hockey life wouldn't even have counted as a bump—and it was like he'd been run over by Scott Stevens. How was it possible that such a *nothing* could be such a *something*, especially for a big, strong guy like he was? The point is: it wasn't possible, it couldn't be possible—and that's why he *knew* that the next doctor, the next treatment, the next *something* would be the answer. And if it wasn't, the one after that would be.

And the doctors saw this the same way he did. It was so clear in their reports. They believed in science, and the best science they knew was incorporated in the tests they administered—and the tests said there was nothing fundamentally wrong with him. So he kept going to see them with a problem that came from some nothing little bump, and they kept telling him he was okay to play, because their science, the best that they knew, told them that there was no reason he couldn't play. The doctors knew there was a lot they didn't know, but they are people of science, and they knew that someday they would know, because *there are answers*. They didn't want to shut him down. They didn't want to tell him he can't play. It was his life, and he wanted to keep playing, and trying—and besides, the answer might be just around the corner.

But now Steve was starting to hear a different message from some of his doctors. One, a "world-renowned Indian doctor," as Vally described him, "told Monty that the damage to his brain was irreparable."

Steve was depressed—because of the setbacks that never seemed to end; because the lockout was over and the season had begun without him; because the end of his career was closing in on him and his life ahead was uncertain. Who wouldn't be depressed?

And he was depressed because he didn't like this new version of himself. He didn't like what he was thinking. He could feel himself getting weaker, and he didn't know what he would do next. This wasn't the same as it was seven years ago, when he checked into rehab and started going to AA. This wasn't even the same as it was a few weeks before, at The Canyon Rehabilitation Center. He was just beginning to realize that AA and rehab centres weren't the answer for the problem he had. He was beginning to realize that he was depressed because *the damage to his brain had caused depression.* Chemicals that had always moved from one part of his brain to another, chemicals that influence mood, behaviour, and function—they were no longer moving the same way, or they weren't being created in the same amounts. Steve's depression was physiological, not just psychological.

Going into rehab at this point was for "maintenance," as he put it, to stop his fall—even if only for a month or two—not for a cure. If there was a cure anywhere, he knew it wasn't with Terry or Pieroth or Kutcher, or even with Carrick or Bailes or The Canyon Center. It wasn't with the best science at that moment. They were all trying, all doing their best, as if their best was the answer. But it wasn't. What they knew couldn't help him. Because what happens when the damage is already done? When it's not about medicine or research or treatment, or about what you can do now? When it's about what you did before? Sometimes treatment *isn't* the answer. Sometimes only prevention is.

What Steve was getting was the equivalent of hospice care—that's all it was—and he didn't know it. He was becoming a different person because he *was* a different person. And he was about to face the biggest test of his life without his best self.

When Vally thinks about those days he spent with Steve in Chicago, and when Steve visited him at his new house only a few months before, he says, "Things went from pure joy to pure misery. This was a completely new Monty."

Vally left Chicago on January 27. The same day, Steve saw Dr. Terry, who wrote in his notes:

[Steve] says that he does feel like he is doing a little better today. He did have some episodes where he noticed some symptoms. These are minor, sporadic and infrequent. He says since he is seeing a therapist currently he is taking antidepressants as well. He says that this seems to be helping. He has been doing some workouts. We have encouraged him again to not work out through symptoms especially, [and] if he has any worsening of any of his symptoms he should stop. He understands this.

On February 1, Steve saw Patrick Becker, the Blackhawks' physical therapist. Becker reported that Steve didn't skate that day but did "his normal routine of cardio instead like he does most days (treadmill, bike, elliptical). Also did about a 30 min. workout in the gym, mixture of strength, balance/vestibular exercises, and flexibility. Monty does express enough sense to shut it down when he doesn't feel good, which he describes as 'spacey' and [having] some trouble with focus."

Steve saw Becker again a few days later: "He skated for the first time. Looked off, as I anticipated. . . . One thing I am trying to get him to do is not go so hard on one day, then to suffer the rest of the day or next, leaving us to take a step back every other day. The largest challenge with him is not overdoing it, as his 'bad days' don't get bad until later in the day, not necessarily during or after the workout."

On February 6, Beckert reported: "Says he got 'dizzy' a few times, but that resolves quickly and did not feel like symptoms escalated too greatly." When Steve saw Dr. Terry a few days later, he told him he was feeling much better, and that the antidepressants had helped him "tremendously." He said he was back doing some informal drills and wanted to start into regular practice, including contact drills. Terry advised him against this.

On February 21, Terry wrote:

> Steve came to the office today and said that he has been
> completely asymptomatic. He has been beginning our
> testing protocol and has done well to this point. He has
> said that he has been under restrictions with no contact
> although he says that he has asked other players to
> undergo some contact and collision activities with him.
> I said that he has done this with the understanding that
> this was against our wishes but he has not had any ill
> effects from them. At this point he is going to re-engage
> in practices. He is going to be on limited contact until
> he is cleared. He understands the game plan.

Later in the month, the *Chicago Tribune* reported that Steve hadn't experienced concussion-related symptoms in weeks, had been skating and taking some contact, and was close to returning to the lineup.

In his March 1 dictation note, Terry wrote: "[Steve] presented to discuss his concussion. He is currently still symptom-free. He has been skating with limited contact with no issues whatsoever." Four days later, Steve's Fitness to Play Determination Form, signed by Dr. Terry, read: "Not disabled. Return to play."

The *Chicago Tribune* reported that even if Steve does begin play with Rockford, the Blackhawks' AHL affiliate, and not with the big team, he "couldn't be happier." He was quoted as saying, "Here we are. I can smile and talk about it a little bit and get on the ice. It's a brand new day."

CHAPTER TWENTY

Dr. Alain Ptito and Dr. Karen Johnston began working together at the Montreal Neurological Institute in 2001. Johnston had been dealing with Montreal Alouettes football players—these athletes had shown symptoms of brain injury but routinely told her that they were fine and wanted to play. Johnston wasn't so sure, and was looking for some way to measure when it was safe for them to return to action. But the players hadn't taken any neuropsychological baseline tests earlier against which any new test might be compared. She also knew that players were conditioned to see past injury—and as with Steve, were very unreliable witnesses. Johnston needed more information.

Ptito, a neuropsychologist, had worked with and assessed several NHL players. He and Johnston decided to conduct studies together using functional magnetic resonance imaging (fMRI) that would allow them to see beyond the basic structure of the brain, which they knew, no matter the technology they used or the injury suffered, almost invariably looks normal. They would put the players into the fMRI, direct them to carry out tasks, and compare the brain activation patterns they saw on the screen to those of normal subjects. Later, McGill University agreed to give a baseline test to every varsity athlete before their seasons began. For Ptito and Johnston, this was the start of their collaboration.

Excited by their work, about four years ago Ptito approached Geoff Molson, CEO and principal owner of the Montreal Canadiens, about doing similar testing with the team's players. By this time, Johnston had relocated to Toronto. When Gary Bettman and Dr. Ruben Echemendia, director of the NHL's Neuropsychological Testing Program and co-chair of the NHL/NHLPA Concussion Working Group, were in Montreal, Molson arranged for the three of them to meet with Ptito. This was a big moment for Ptito. He, Johnston, and others had spent years seeking better answers for their patients. When some new idea or direction seemed promising, they had fought for what had become increasingly scarce research grants in order to know more and know sooner. Now Ptito had some encouraging new approaches and he had in front of him the commissioner of the NHL, the most important person in the most important league of the most popular sport in the country, who so obviously had a fundamental stake in the future health of the league's players; also a fellow scientist who appreciated the potential impact of his research; and the owner of a team with a history of doing the right thing, whose family had a tradition of giving to hospitals and educational institutions.

"We showed them some pictures and images, and told them what we were doing. And they responded well," Ptito remembers. "And the fact they were coming to see us seemed to show that they were open to listening." Ptito offered some ideas for research that the league might support, but more importantly he said to them that he thought this was the right moment for a pilot project, and that the Canadiens were the ideal team to be involved in it.

"The NHL has the money to be able to sustain this kind of research," he remembers thinking. "They could change things, and they could say, 'Of all the teams, let's start with the Montreal Canadiens.' The whole team could be scanned before the season started, and if something happened to any of the players we could retest them, and compare, and see how they're doing. That would be fabulous." He was excited at the prospect and what it might mean for the future. "Because then, I knew, we could avoid a lot of recurrence of concussions."

Ptito also understood the limitations of the basic neuropsychological test that teams were using, and so did Echemendia. "The players are so familiar with it," Ptito explains, "that at baseline they can 'fake bad,' so that if they get a concussion and do the test again, no matter how badly they do, it will come out favourably to the way they were at baseline." Athletes with a concussion, Ptito adds, "have a tendency to 'fake good' in the later test because they want to return to play. They want to keep on doing the things they love to do. It's different from somebody who's been in a car accident, where they're in litigation, they try to 'fake bad.'" Baseline fMRIs, Ptito believed, could help the league avoid the problem of wrong diagnoses.

"But nothing came of it, really," Ptito says of the meeting with Bettman, Echemendia, and Molson in 2013. "We never heard from them again." Maybe the league believed that other projects were more promising. Ptito heard indirectly that cost might have been a factor. "To do an fMRI, with analysis, would [cost], I don't know, about $2,000 each time. But are we really talking about a huge amount of money? With the money that is going around in the league, I'm not so sure. And if you can prolong a player's career? It's like they want the players to stop when they're thirty. Because we're seeing more concussions, and with their cumulative effects the players lose their ability earlier in their career."

But Ptito thinks there is a more basic problem, too. "People are afraid of change. Why try to fix something that doesn't seem to be broken? It looks good, what the NHL is doing, and there has been an improvement, and as they will tell you, 'We're doing neuropsych testing at baseline, and we're repeating it. We're taking the players out of the game.' And yes, this is a good thing. But certainly it's not enough." To Ptito, it comes down to one point: "I don't know to what extent they're committed to saying, 'We have to help the individual.'"

Ptito has to believe that Bettman and the others are committed. Ptito is a researcher and a clinician. He is also human. He knows the value of what he and others like him are doing. He knows the consequences of brain injury; he sees it in his patients. He has to be hopeful. And he knows that Bettman, Echemendia, and Molson were listening. He knows they are sensitive to the problem because he met them and saw them up close that day, and how could they not be? He knows that listening is not just tilting your head and furrowing your brow in a certain way, that listening is responding, and sensitivity is responding with action of a magnitude similar to the magnitude of the problem. Otherwise, both are just pretend. This isn't about *doing* something, he knows. It is about *achieving* something. But every so often he is drawn away from his hopefulness, and he sees what isn't being done. Then a darkness creeps in.

"Do you think I'm jumping up and down with happiness?" he says when he talks about the possibilities missed. "I've been working on the brain for over thirty years. I'm very frustrated. I know I have some answers, yet I cannot implement them. But you know,

I try to control my emotions because if you don't, you get even worse results. So when I'm asked to give interviews, I never refuse. I go to conferences. I present these results. I try to communicate everything that I know to anybody that invites me to speak. That's what I do."

When not much changes, first he thinks it's *their* fault—those in authority who have in their hands the power to change things—then, like Johnston, he thinks it's his, because he hasn't delivered his message the right way. He tries to imagine how *they*, how all of his audiences, think. "The league has a business interest," he says, "so maybe if I talk about things in a business way, that if they do these scans before and after, if they prolong an individual's career then it may be in their own best interest in the long term; that there's a financial dividend." But mostly, Ptito believes, the answer for his patients, for anyone with brain injuries, lies in knowing more, so that others can know more and make smarter decisions. It has to be this way. Knowing more is what he does. Making smarter decisions is what Bettman, Echemendia, and Molson do. They are all linked.

Ptito talks excitedly about new technologies that will allow scientists to know more. "I just got a grant to buy a new MRI," he says. "The one we have now doesn't let us see any change in the brain's structure, and we know that there must be changes, particularly in connectivity. It may not be that one single area of the brain is affected by an injury—it's not that you're going to see a blob of blood or anything. It may be that the connections between two regions are interrupted." He and his colleagues will have the new MRI in 2018. They also have almost enough money now to run a positron emission tomography (PET) study on retired players. A radioactive substance that binds to certain proteins will be injected into the players, and a subsequent PET scan might reveal the presence of CTE. This, while a person is still alive.

"Imagine," Ptito says animatedly, "then we can start working on medications. We can repeat the tests and see if there is improvement. We can prepare the families, and explain to them what may be happening because we've seen this and that. We can prepare the player. We can give him psychological support. Imagine somebody like Steve Montador, if you're thirty-five years old and you receive a diagnosis of CTE. What do you do? There's a whole life ahead.

"You have to help these people. Are they going to remain depressed for the rest of their lives without anybody doing anything? It's not everybody that is able to commit suicide. We would know the CTE is there. Perhaps we can improve their quality of life and slow down the process." To a scientist, this represents hope; to a thirty-five-year-old hockey player, it does not. Ptito doesn't need these new machines to know that his patients have a problem. He sees it in their behaviour. They have memory loss that is a whole lot worse than simply forgetting where they put their phone or their car keys. They make crazy, erratic decisions that are a whole lot worse than just having a bad day. They are anxious and depressed, and some things set them off when they never did before, and other things don't when they did and should. There is something wrong with them; he can see that. It doesn't matter what his fMRI or PET scan shows or doesn't show.

But he needs these machines to also find out lots of other important information: to see if CTE is like other diseases that are better known and better studied—Alzheimer's, Parkinson's, ALS—so that through these other diseases he can learn, postulate, go off in promising and exciting directions.

Ptito also needs these machines so that other people, the deci-sion-makers and those with influence who aren't scientists, can see the damage for themselves, and understand the problem. When they see a broken bone on an X-ray, they know. When they see a tumour on an MRI, they know. But at the moment they can't see CTE on any screen. "Where is it?" they say. "Show me. If I can't see it, how do I know it's there? How do I know if anything is even wrong? He looks fine."

Henry VIII, in the early years of his reign, was considered even-tempered and wise. After two jousting accidents, one that left him unconscious for two hours, he was erratic and rage-filled. A Yale University neurologist thinks there may be a connection between the incident and the change in his behaviour. But Henry looked fine—and kings can do anything and be forgiven (except, perhaps, by the Catholic Church).

In recent years, the concussion debate has become all about CTE. And before death, CTE isn't there because we can't see it. After death, it's too late. With the focus on CTE but with no smok-ing gun, all the signs seem not so important. They aren't dramatic enough. But to Ptito and others, something *is* wrong, because depression is there, because crazy, stupid decisions are there, because memory loss is there—and these aren't there in a normal, healthy thirty-five-year-old male. Brain injuries are *not* invisible. This person, that person may look fine, but hang around them a bit. Watch them as they live out their day. Do they still look fine?

Ptito is a clinical neuropsychologist; so is Echemendia. Neuropsychologists study people and behaviour. They can't see everything that has gone wrong in the brain, but they don't need to. They see a problem with their own eyes, and in the results of the tests they administer. They don't see it in the concrete brain; they see it in the concrete person. To others, who don't wish to see unless they want to, that isn't enough.

We all want to *know* things for certain. And we especially want to know if the consequence of our knowing would mean the need to change what we already do. Change is a nuisance. It takes time, costs money, and creates uncertainty. We already know how to do what we are doing, but we might not be able to learn something new. We know the good of what we already have, and we know how to live with the bad. And we only *think* we know the good of the new; and we certainly don't know the full potential of its bad, or know if we can deal with that. So if depression, erratic behaviour, memory loss, or worse have to exist, we don't want them to be because of thunderous hits or uncounted blows to the head. We want them to be about things that we can't do much about, like genetics; or things that we think we have some control over, like alcohol or drug use. So we do what we want to do, and make its case, not what needs to be done.

Many players and non-players who have suffered head injuries have substance abuse problems. They have pain, and they medicate their pain through doctors—legally or illegally, correctly or incorrectly—or they medicate it themselves. After a while, pain is pain. You get rid of it any way you can. When a player who has had head injuries and also a history of alcohol and drug abuse shows symptoms of brain damage, what's the story? Is it about hits, or addiction? It seems so confusing, so complicated, so unclear. It's the chicken and the egg.

It seems easier, even more responsible, not to think about having answers at all. There is so much we don't know, and shouldn't pretend to know. Shouldn't we just wait? But our perceptions of what is wrong and why don't wait, with unfair and destructive consequences. Because one thing we *can* see; one thing we can't. We can see the unhealthy symptoms—the drinking and drugging. We can't see inside the brain. So the story, about hits and addiction, comes to be about addiction first, because it is easier to understand and see. (And hits get relegated to a parenthesis.)

And the story of the brain-injured player can be so sympathetically told. It's about this really good guy who loved to play, who loved his teammates, and loved the game. Who played the game the way it is supposed to be played—with dedication and respect. He stood up for his teammates no matter the circumstances. He stood up to his opponents no matter how big and tough they were. He worked hard on the ice and off. Injured or sick, he played because his team needed him, and his teammates counted on him because they knew that they could.

But he was young, and things happen when you are young. Young people are healthy and strong, and think they are invincible. They want to do everything; Steve wanted to do everything. He had money. He pushed the limits, because he was curious and young; because he was a player. Youth, money, opportunity—it can be a dangerous mix. And Steve was getting older. He could see the end of the line. He had played hockey all of his life: it's what he did, it's what he was about, it's where he found his purpose, his meaning, his pleasures and joys. Now his hockey life was about to be over. Now he had fifty years of his life in front of him. What was he going to do? Where would he find his new purpose, his new meaning, his new joys? He was feeling the pains of injury and age. Of course he was. He was depressed. Who wouldn't be? He was lost.

It is all so understandable and sad. For everybody involved, the brain injury narrative can be what you want it to be. For players and former players, it can be about alcohol, drugs, and "transition," because if it were about CTE that would haunt them every day of their life; and alcohol, drugs, and transition, in theory at least, they can control. For the league, it can be about something rare and tragic. It can be a one-off: the unknowable life of one person. In this way it doesn't have to be about changes in the game, about its more frequent, more forceful hits; about decisions made, or not made. In this story, there are no villains. Only the ineluctable circumstances of life. And a really great guy.

Thank you for your service.

Just when Dr. Ptito thinks that things are getting a little better, something happens that makes him wonder. He remembers a game in December 2015 between Montreal and Columbus. Canadiens forward Tomáš Fleischmann was skating up the boards with the puck, and Nick Foligno of the Blue Jackets stuck out his knee to stop him. Nathan Beaulieu of the Canadiens confronted Foligno, and the two of them squared off to fight. Both are strong, tough guys, and things were fairly even until Foligno hit Beaulieu with a right hand to the head, and Beaulieu crumpled to the ice. He was then directed to a "distraction-free environment,"—a "quiet room"—as required by the NHL's regulations, in the Canadiens' dressing room area, where he went through the league's mandatory concussion testing and protocol administered by the Canadiens' team doctor.

Seven months earlier, during the playoffs, Beaulieu had been blasted with a shoulder to the head by the Senators' Erik Karlsson, a hit that was run and rerun on highlight packages all over Canada and the U.S. Ptito was at that game. He also watched the Columbus game on TV. "Beaulieu got up and he was kind of wobbly [after the Foligno punch]," Ptito says, "and to me it was a concussion for sure. Then a few minutes later, he returned to play."

Ptito was incredulous. It seemed to him an easy case: Beaulieu was in a fight, everybody was watching, nobody was distracted from it by anything else; everybody saw the punch, everybody saw him go down, everybody saw him wobbly; they've all read the stuff about head injuries and concussions. It was the twenty-sixth game of the season, not a playoff game; there were fifty-six more to play.

So, Ptito knew, Beaulieu was not coming back to play—*and then he came back to play!* How could that be? Then Ptito answers his own question: "They tested him in the locker room and he passed." Period. But he knows, and the Canadiens' doctors know, that players find ways of passing these tests because they *want* to pass these tests. Beaulieu is a big young guy. Known mostly for his play with the puck, he has carved out a role with a reputation for standing up for his teammates—one that matters to him and to the team. That's why he took on Foligno. Yet not many months earlier, he'd been crushed by Karlsson and didn't play for seven games. That can't happen to team guys. Team guys have to stay in the lineup. They have to be there to stand up at stand-up times, and they can't do that taking a test in the dressing room. The doctors knew that. They have their tests, but they have their eyes too, and they knew Beaulieu's history. That's what mystifies Ptito. Why didn't they say, "I don't care what the tests say. You're done for the night, Nathan."

It is the power of wanting to play. A player *absolutely* wants to play. One, because he wants to play. Two, because he thinks his team needs him. Three, because his teammates expect it of him. Four, because his coach expects it of him. Five, because the fans hope it of him. Six, because every player is replaceable, and if he doesn't play, like Wally Pipp he may never play again. Seven, because the commentators will say, "He's a hockey player. He'll be back." Eight, because he is young and thinks he can, and he thinks there are no consequences if he does. Nine, because if he doesn't play because of a head injury, he will become a more vulnerable, less desirable commodity. Ten, because if he does come back to play in the same game, his injury can't be a concussion, because if it *were* a concussion he wouldn't be allowed to play in that game (even if it is one).

It is not just about adults—players—being old enough to make decisions for themselves. It is about these adults having people around them who are more than just notionally sensitive, notionally listening, notionally good guys, notionally with a reputation for being among the best in the league; people who will stand up to the player's drives and needs that they helped to instill and create and encourage; people who will safeguard a player from himself. Tests can reveal; and tests can also protect. Tests can be a light; and tests can also be a shield.

For leagues reluctant to change, CTE is the bad news. It is that awful condition whose symptoms strike forty years before their time and make a person someone else. It is that terrible condition about which something simply must be done so that other players, and their families, don't have to go through it, too. As a result, CTE has become the focus of the sports head-injury debate. But CTE is rare—or if it isn't, we don't know that it isn't, and we won't know that until a lot more players die and their brains are examined. So, for hockey players or football players, the problem—at worst, as far as we know—is a rare problem: sad, tragic, but a problem of relative one-offs.

For the unlucky ones and their families, the leagues will do more, whether of their own volition or nudged along by the pressure of class action lawsuits. The leagues will help those who suffer from CTE financially, and some day perhaps with medication and treatment, guidance and therapy as well, so that the athlete and their family can live the next forty years of life better. In time, perhaps, CTE might be turned from an urgent condition to a chronic one. Much as retired players today deal routinely with chronic knee and hip problems, one day they might deal with chronic depression and memory loss as a chronic brain problem.

The focus on CTE has shifted attention away from sub-CTE conditions such as depression, memory loss, and irrational behaviour, much as a knockout in boxing or an obliterating hit in hockey shifts attention away from frequent, sub-concussive, wear-and-tear blows to the head. CTE the red flag has become CTE the shield.

As a scientist, Ptito can study, he can learn, he can tell others what he has learned, but he is not a decision-maker. He cannot prevent. Only others can do that. As a clinician, one on one with his patients, he has more influence and effect. He cannot prevent their initial injury, but he can help to prevent their next one, and he can advise patients to stop their risky behaviour because the next one might be worse. Like Johnston, he finds this side of his work immensely satisfying. But to do his best for his patients, he knows that he has to be effective in terms of the big picture, too. Research and technology are expensive. Funders want big-picture payoffs; they want lots of people to be helped. He needs the NHL to do more because NHL players need to do better, and his patients do, too.

So he and Johnston keep telling their story.

CHAPTER TWENTY-ONE

Steve had not played for almost a year. Chicago could now send him to Rockford for up to fourteen days on a "conditioning loan," to get him into playing shape. Or they could put him on waivers, which would allow every other team twenty-four hours to claim him, and his contract, with no compensation to the Blackhawks in return. If Steve were in Rockford, or on the roster of another NHL team, Chicago would get some relief under the league's salary cap, yet this late in the season they would have no opportunity to use that relief to acquire another player for their playoff run. Given Steve's injury, his age, his contract, and because the Blackhawks no longer thought he could help them, they put him on waivers. No other team claimed him.

Now Steve had his own choice. He could see the demotion to Rockford as one more blow in a season full of very big blows, and put front and centre in his mind the headaches and depression he was experiencing so that nothing else in his life mattered. Or he could still pick up his same paycheque for the rest of the season and for two more seasons—not risk further injury, and not subject himself further to this psyche-crushing, soul-crushing cycle of hope and despair.

He could go easy and feel lousy. Or he could go hard and feel lousy. To Steve, why not go hard? Why not play? Why not hope?

Steve made his debut for the Rockford IceHogs on March 15. It was his first AHL game in more than nine years. He thought he could help the IceHogs in the way he had always helped his team: being aggressive, pushing, forcing, never letting up. But not having played in almost a year, he wasn't so sure of what he could do and what he couldn't. He didn't have the same feel for the game, or for himself. But he *was* certain he could help the team off the ice, as the experienced veteran who had played more than 500 games in "the Show." He had done what his new teammates had only dreamed of doing.

With the IceHogs, he would arrive early for practice and stay late, he would work hard and train hard, he would put all of himself into what he was doing, as if he wasn't sorry for a second that he was where he was—and not where he wasn't. And if he did, what right would his new, young teammates have not to do the same themselves?

After practice, he went to lunch with them. Often he bought; after all, he was the one with the big NHL contract. Sometimes he took them to dinner. He told them again and again what he knew and what they could only hope—that the NHL was not that far away. That, sure, there were the Toews and Kanes and Keiths, the franchise-defining players, but there wasn't much difference between the rest of the team and them. "Look at me," he said to them. Sometimes in the NHL he'd been a 3–4 defenceman, usually a 5–6, sometimes a 7; always on the verge of being an 8 or 9 and out of the league. His IceHog teammates needed to know that they were amazingly, shockingly close. They had only to keep working, preparing, getting better for tomorrow by focusing only on today, because someday on the big team there would be an injury, a trade, a cap problem, and they would get their chance. But they had to be ready, because if they weren't, that chance would pass to somebody else who wasn't much different from them either, and who *was* ready.

Steve talked to his young teammates as if they mattered, and listened to them as if they had something to say. He had his own down moments—how could he not?—but never for long. "He knew he had a purpose," Missy explains, "and he thought, 'This is my purpose now. To help these kids.'" As Steve had with every team he played on, he did what needed to be done.

He was a player, not an *NHL* player or a *veteran* player, and his IceHog teammates were players, too. Less than a week after he arrived, he did an interview with the IceHogs Broadcasting Network. It was after practice in a poorly lit corridor near the ice; Steve was still wearing his jersey. The interviewer, in several different ways, asked him what it was like—to be playing again; to be back in the AHL. Steve said, "It's a real blessing to be down here," and that he was "definitely rusty," and glad to have played his first two games on back-to-back nights so that he didn't have time to worry too much about the challenges of returning to the game. On the ice, he said, he was still thinking about "where I'm supposed to be" rather than just being there. But while this community-station interview in Rockford, Illinois, was a long way from his scrum with the national media just outside Toronto the previous summer, he sounded upbeat. He looked at the interviewer the same way, listened the same way, paused before he spoke, and answered the same way.

The IceHogs had started the year slowly, improved in the second half of the season, and were still many points out of a playoff spot when Steve arrived. He played fourteen of the team's final seventeen games, in which Rockford went 12–4–1, winning eight of their last nine games, and missed the playoffs by only two points. Steve got into one fight. The circumstances were predictable. The IceHogs were on the road, playing against their in-state rival, the Peoria Rivermen. Behind 3–2 in the second period, Steve went after Stefan Della Rovere of the Rivermen, they shoved, Della Rovere knocked off Steve's helmet, then the two of them mostly held on to each other before the linesmen intervened. The IceHogs came back and won the game, 5–4. Steve was willing to do what players do and take the risk, even in a minor league that would take him nowhere.

When the Rockford season ended, Steve and seven of his IceHog teammates were brought up by the Blackhawks. The spares on every hockey team, those added for the hoped-for long, injury-laden slog of the playoffs, are known as the "Black Aces." The legend goes that when "Wild Bill" Hickok was murdered playing poker he was holding two black eights and two black aces, what came to be known as the "dead man's hand." The Black Aces are a hockey team's "dead." Their unwanted. They practice in the same arena as the big team, but they have their own dressing room and practice time, lest they "infect" the regulars. And they love everything about being there. They get the taste of an NHL city during the playoffs, the feeling of excitement and importance that is everywhere, all the while dreaming that with the manager and coaches of the big team looking on, they will be spotted, rushed off the practice ice, and into the starting lineup for the series-deciding game, with only fabulously fulfilled dreams ahead.

Steve organized the golf games before the Black Aces' practices, just as he had for Crosby and Tavares at Vail. He was the unofficial captain of the Black Aces. When the Blackhawks were in Minnesota for their opening-round series against the Wild, Steve, being back in Chicago, invited the guys over to his place, where they ordered in food and watched the games. Chicago beat the Wild, then the Red Wings, and Kings, finally winning the Cup against the Bruins. The team played twenty-three games; only one of the Black Aces, winger Ben Smith, played at all: 10:23 minutes in the third game of the final series. By that time, Steve was gone.

Mike Keating remembers the day. He had left Peterborough earlier in the week and was driving to Chicago. "Hey, I'm coming to see you," he said to Steve. The Blackhawks were playing the Red Wings and were down three games to one in their series. "And Monty says, 'Hope you brought a suit.'" Keats was too excited to ask him why. "So I get there, we're at the arena and he's got me walking in with all the players, going through the back corridors and up into the stands. He treated me like I was one of the guys. I ended up sitting with him on one side and [retired Hall of Fame defenceman] Chris Chelios on the other. That night he found out he was being sent home."

The Blackhawks were in the midst of a tough series against Detroit. They hoped to have two more rounds of playoffs ahead. Steve had come back late in the year, played when he felt like crap, fought, almost helped the IceHogs into the playoffs, and was a Black Ace for the first time in his career. The day Keats arrived, a few of the other Black Aces had been told to join the big team, to practice with them and travel on the road. The rest were told they weren't needed. Steve was one of them.

"When that happened," Keats recalls, "it broke him a little bit."

The next day, Steve did his exit examination with Dr. Terry, who reported "no complaints, symptoms or issues re: concussion."

He listed Steve as "Not disabled" on his Fitness to Play Determination Form.

Later in the day, Steve went out with Keats and some of the other now-former Black Aces. Keats ordered a glass of wine, Steve reached over and drank half of it. Keats was stunned. "Hey, c'mon man," he said, "I don't agree with this." A short time later, Steve sent a text to Missy. "I'm drinking," it said.

"That was the day that changed everything," she says. His drink with Keats and the guys turned into a ten-day binge.

The Blackhawks won the Cup; Steve had a four-year contract with the team, but Steve didn't win the Cup. A team goes through things together; it loses and wins together. The players know who is a member of the team and who isn't, and that season Steve was a member of the Rockford IceHogs. Daniel Carcillo had played four games for the Blackhawks in the playoffs; he was a Stanley Cup winner. Steve tried to be angry about how he felt. He tried to feel sorry for himself. He just felt one more sadness.

A few days after his binge, he checked back into rehab. For the next twenty-one months, he would have some good days—and many more bad ones.

On June 27, three days after Chicago had won the Stanley Cup, general manager Stan Bowman announced that the team would be using one of its two compliance buyouts on Steve. With the Cup win, many of the Blackhawks players had earned bigger contracts, and the league's salary cap would not be rising sufficiently to meet these new demands. By unloading Steve's contract, the team would save $2.75 million in cap space in each of the next two seasons. Compliance buyouts had been agreed to during the last CBA negotiations—the ones in which Steve had played a part during his dark days at The Canyon Center. He would get his money, but at his age and in his state of health, he needed a contract to protect him, and that was gone. What he had helped to negotiate had ensured the end of his NHL career.

His days in rehab over, Steve felt well enough to become hopeful again. He went to see Dr. Kutcher in Ann Arbor, and Dr. Carrick in Atlanta. When he got back he was all excited. "I'm good," he told his friends. "My head is clear." He was proud of himself. He was working out again; he was going to AA meetings; he was sober. He had a plan. Kurt Overhardt, his agent, started calling around to NHL teams. There were no takers. But Steve was undiscouraged. He hadn't really played for a year and a half, he said to himself. He'd have to prove himself again. But he was feeling better, and he got his buyout; if he had to play in Eastern Europe in the KHL, then that was the next best league, and it didn't have to be forever. He would get in a full year of play, and then he would only be thirty-four, still lots of time for a 5–6 defenceman. He might get another shot. Then he'd know, he'd know *for sure*, if he could still play. He didn't want to go out this way. He had come into the league as an undrafted free agent; he would go out on his own terms. Besides, playing in the KHL would be an adventure.

On August 11, 2013, he signed with Medvescak Zagreb. "Steve is a defender who played 10 years in the NHL and [in] four seasons he had more than 20 points per season," Aaron Fox, Zagreb's "athletic director," said on the team's website. "[He] has fully recovered from a concussion, is completely healthy . . . [and] will bring added stability to our blue line."

Zagreb is Croatia's capital and biggest city, with a population of almost one million. Medvescak had previously been in the Austrian League, and was one of only three clubs in the twenty-eight-team KHL located outside the old Soviet Republics. The other two were in two traditional hockey countries: the Czech Republic and Slovakia. The KHL season began in early September. Medvescak, as a new team, would need to be put together quickly; Steve had to get there for training almost right away. Among his new teammates were Jonathan Cheechoo, who in 2005–06 with San Jose, had led the NHL in goals with 56, and Matt Murley, who Steve had played against in the Quebec peewee tournament twenty years earlier.

Zagreb would finish sixth out of the fourteen teams in the league's Western Conference—ahead of CSKA, the legendary Red Army team of Moscow—and make the playoffs. But the road trips were long and punishing, drinking came easy for Steve and the others, and Steve's concussion symptoms wouldn't go away. After eleven games, he came home.

He had been bought out by the Blackhawks, received no offers from other NHL teams, and had left the KHL behind. He had no place left to go. His phone still might ring, and he would keep on training in case it did, but this time Steve knew he was done. Until this moment, everything had been about his career. Now it was about his life.

He still had his place in Chicago, or he could go back to Toronto, but now without a team he had no real home. George Parros, his former Anaheim Ducks teammate, was playing in Montreal but had kept his condo in Hermosa Beach, and Steve rented it from him. One of a string of beach communities south of Santa Monica, Hermosa Beach was affluent, mostly young, with lots of cafés, bars, and beach-shack restaurants. Its biggest employer was 24 Hour Fitness. It was also a place for serious surfers, paddleboarders, and beach volleyball players. It was a place for Steve to get some distance, thousands of miles away from his friends and family, from his old life, from too many people who would ask him too many questions. He had some thinking to do. He had to get himself back together and sort out what he would do next. With its sun and sea, Hermosa Beach was a place for him to get healthy.

He had no practices to get up for or games to play. He didn't have teammates with needs he had to meet, or coaches he had to please. He didn't have fresh injuries to heal. He could hang out at Starbucks with his laptop and live a summer life all year round. Nobody knew him here; he had no expectations to live up to. He had never liked to tell people he met that he was a hockey player. He certainly didn't need to tell anyone now. He was single. He had money. He could put together the pieces of his life however he wanted them to fit.

He could also be alone. Nobody could just drop in. He could text or call his family or friends when *he* wanted, whenever he had something to say and was ready to say it, and everybody would be thrilled to hear from him. He could do it on "good brain days," as Missy describes them, when he was sharp, and up, and able to sound just like Monty, and whoever he texted or called, he knew, would text or call others, and they all would say, "Monty sounds great. He's surfing and paddleboarding. He's in his flip-flops and doing a million things," and they would all laugh and know he was all right because this was the Monty they knew. All of which would give him even greater freedom to do what he wanted to do, and to hide.

But the dark clouds wouldn't go away.

Just before Christmas 2013, Nick Robinson, one of his Peterborough buddies, visited him. "Monty picked me up at the airport," Nick recalls. "He was coming from this hypnotherapist and he had lost his wallet. So we spent the next hour and a half retracing his steps, going to all the places he had been that morning and afternoon. Then he gets this call, and the wallet's at the hypnotherapist's. That was my first experience with his memory problems."

Nick wasn't entirely surprised to find out that Steve had been drinking again. He hadn't heard from him much in the previous few months, which was unusual. Nick had had a problem with drinking himself when he was younger, and had used Steve as his inspiration and changed his life. Now he feared Steve was turning away from those he needed most, right at the time when he most needed them—because he was embarrassed, because he didn't want anybody to see this Steve.

But in the few days he was with him, Nick also realized Steve's problem wasn't just alcohol. "He was taking medication for depression and anxiety. He was having convulsions from all his concussions. I was also concerned about the decisions he was making, and the way he was talking and thinking. Monty was always a very deep thinker," Nick says. "He would speak logically. He would speak effectively. He would speak articulately. And he was a very good listener. That's why the dialogue between us was always so great. He would talk; I would talk; it was always fifty-fifty. Now he talked and talked and wouldn't listen." After a few days, Nick had to return to New York to go back to work.

Steve had no continuity in his life. No co-workers who saw him every day. He had nobody to return home to, no wife, no partner, no roommate who might notice something. Steve was on his own, just the way he wanted it.

A few weeks after his visit, Nick got a call from Steve. "I'll never forget it. Monty says, 'I'm going to follow your lead, chum. I am going back into the program.' It was a pretty fantastic moment," says Nick, "because I felt there was hope."

Steve still saw Daniel Carcillo occasionally. Carcillo had been traded to the Kings that summer, and the two of them spent Christmas together. "I could tell that something was up with his head," Carcillo recalls. "He was always forgetting his phone or his keys, and he didn't really look you in the eye when he was speaking to you, or hold your gaze like he used to. It just looked like he was hurting." Carcillo also knew what Steve was doing in all the hours he wasn't biking or training or dreaming of what he might do next; what he was doing at Starbucks with his laptop. "He was researching concussions," Carcillo says, "and what the fuck was going to happen to him because of them. And the more knowledge he gained, it seemed the worse he got, realizing maybe that he wasn't going to reverse the symptoms and the memory loss and the headaches." Carcillo had sustained several concussions himself. He knew that Steve was now doing a million different things because he loved to do a million different things, but also because he needed to keep his mind racing so he didn't feel his headaches or his anxiety or his depression or anything about anybody, especially himself. But a million things weren't always enough.

And when they weren't enough, Steve drank; and when he drank two, he drank ten. For more than seven years he had fought those battles and won. Now he was losing. He no longer had a life-structure around him; the incentives of his career and of money were gone and not coming back, and he knew that, and he hated that. But he had so many other lives to live, and a whole big world to discover and learn about, and new people, and all the time and money he needed to do that. But what he didn't have, what was gone and not coming back—and what he hated even more and couldn't live without—was himself. He could still trot out the old Monty (or what he thought others thought the old Monty was, the one they

289

loved), but not very often. And that Monty wasn't Steve now. That Monty was gone.

"He knew what was going on with him," Carcillo says. "He had all those extra sets of keys made. You don't do that if you don't know something."

Steve tried to focus on what he might do next. He was good looking, he spoke well, he was smart and funny, he could be outrageous; players liked him and would talk to him. He could do something in broadcasting with one of the NHL teams or with the NHL Network. Or something with the NHLPA. He was a player through and through, and had a player's mentality. He had been involved in two CBA negotiations; he was active and committed; he worked hard and did his homework. In the midst of his suicidal thoughts in the rehab centre in California, he had helped to get the CBA deal done. The NHLPA could be his new team, and he could help all those players in the league who were going through what he was.

Or he could help players from the team side of things. He hadn't been drafted; he'd had to work his way into the league. He'd always had to improve, and to prove himself. He was a good teammate, and he had been through so much. Teams were getting younger. He had always worked well with young guys: he related to them, and they related to him. He could get involved in a team's player development, for example. Craig Button, his old GM in Calgary, even imagined something bigger for him. Steve knew everything about a team and how it operated—not only about players and coaches, but about managers, owners, scouts, trainers, fitness people, and the guys who cleaned the arena—because he talked to everyone, and listened to everyone. Button thought Steve might run an NHL team someday.

There was also the possibility of doing something outside of hockey. He was a part-owner of Andy O'Brien's gym in Etobicoke. The two of them had talked about what they might do together: O'Brien, the fitness guru; Steve, the spirit of a gym. Or they might do something different and new. A few years earlier, when he was living in Calgary, O'Brien had turned a house he rented into what he called a "Bed & Barbell," with a gym in the basement, a yoga area, and an office suite where he did presentations on training and nutrition. "It was designed so that the athletes could stay with us for a couple of days and experience what it's like to eat and train a certain way." Steve had visited him and loved what he saw. "We even had organic bedsheets," O'Brien recalls, "which Monty thought was the coolest thing." Someday, the two of them might do something like this together, only bigger.

Or he could help out with different charities. He had once gone to Tanzania with the Right to Play organization, along with his Flames and Bruins teammate Andrew Ference. And if not Right to Play, then some other organization or group where he could help people. "He was always the first to care about something," O'Brien says. "I really believed that his hockey career would have been the least of his accomplishments."

And, of course, Steve being Steve, there would be some fliers in his future. "He was always the locker-room guy with the next big idea," Vally says with a laugh. A clinic with one of Carrick's gyroscopes; an online poker site in Russia. "He had this guy who was able to take money from North America, apparently legally, through a wired account and funnel it to Russia, because the headquarters were there. Monty called me and wanted me to invest in it. I said, 'Monty, I don't have the money,' and he said, 'That's OK, I'll front it for you, chum.' I get off the phone and tell my wife. She's like, "He's crazy.' I said, 'Yep, he is.'"

Steve made calls on his good days in Hermosa Beach, but more often, unable to escape his present to think about his future, he was on his laptop, reading about all the products and treatments that might make his head better, all the ways of looking at life that might make him a better person. One time he called Vally and told him he had joined a group for sex addicts. Vally told him he wasn't a sex addict, and accused him of being addicted to rehab groups. "But, Vally, Vally," Steve said, "if I abstain from sex for eighteen months, then it's going to be the most amazing experience, and I'll get a better understanding of which girl is right for me." Vally recalls, "He was months into it when he stopped."

Steve also joined various men's organizations. One called itself MDI, Mentor Discover Inspire. Its mission: "To cause greatness by mentoring men to live with excellence and, as mature masculine leaders, create successful families, careers and communities." Another was called Full Spectrum Man, created by David Fabricius, a self-described "world renowned leadership, sales-motivation and life optimization speaker resource, a master executive coach, Fire-Walk facilitator and specialist trainer." Its members went on two-day, sunrise-to-sunset, intense mind-body-spirit training adventures. Steve liked the way the men talked, the meaning that they sought, the way they cared for each other. They didn't know what Steve had done before or who he had been. They only knew what they heard when he spoke.

Steve wrote about these groups in his journal, about their principles and purposes, and why they mattered to him. But by this time, late winter 2014, his writing is hard to decipher, his thoughts were often incomplete and difficult to comprehend. Later his entries grew even more frenetic, then got better, then got worse, until they almost stopped.

Steve hadn't seen his family or most of his buddies for months. But during one email exchange, his brother Chris had suggested that they create an annual tradition, a brothers' trip to a Stanley Cup Final game. The Rangers had advanced and were going to play the Kings in the 2014 final. The two of them could meet up in New York. Vally, who was now working for the Rangers, could arrange the tickets. They might even see Nick. Early in June 2014, Steve flew to Toronto on his way to New York.

He called Vally from Toronto; Vally told him he'd pick him up at the airport. "Then he says to me, 'I am going to fly private.' And I say, 'Monty, private is 8,000 dollars . . . 250 bucks can get you to Hartford.'" Vally pauses. "He flies private. So he arrives and comes into my place like a bull in a china shop, right away to the fridge, grabbing beers, throwing them back. My mother-in-law, who loves him, stayed up all night with him, trying to talk to him. I woke up in the morning and he was out on the beach with a dozen beers around him in a lawn chair, on the phone. He hadn't slept. We had a day planned where we were going to feed the kids and go paddle-boarding. Now he had to sleep." The next day was no better. "He's on the phone, and I'm like, 'Monty, what are you doing?' And he says, 'I'm trying to get a helicopter to go to the city for the game. I don't want to get stuck in traffic.' And I'm like 'Monty, where do you plan to land the helicopter?' He says, 'On the beach.' And I'm like, 'Monty, what the hell are you doing, man?' In the end, he took a cab, but he was rude. He was angry. He couldn't hold a thought." The weekend was a disaster.

It was the last time Vally saw Steve.

About this time, Steve met Chantelle Robidoux. Chantelle was from Buffalo but had moved to L.A. after high school, more than a decade earlier. She had two friends back home who had married former Sabres players—and teammates of Steve—Rhett Warrener and Sean O'Donnell. Her friends introduced Chantelle to Steve in L.A. A few months later, Chantelle told Steve she was pregnant. Steve, with a hundred thoughts in his head, now had one more.

In late August, Mike Keating visited Steve in L.A. Keats was excited to see Steve; Steve was not excited to see him. "He just wanted to be as far away from everybody as possible," Keats recalls. "I'd say, 'Hey, how are you doing?' And the first thing out of his mouth was, 'I don't want to talk about it. Everyone's all over me.'" On this trip, Keats says, "I started to realize this was much more serious than I thought."

Steve told Keats about the "men's groups," and about how they were going to open his mind. "He thought if he couldn't fix what was already broken, maybe he could open up more of what was already there." He told Keats about all the other things he was taking and doing. "He had these pills that stop your anxiety. He had pills that stop your depression. I said to him, 'We know rehab doesn't work for you. We know AA doesn't work. Obviously, these mentors don't work either. What are you looking for?'"

Steve also told Keats about Chantelle and the baby. Steve didn't know what to do, Keats remembers. He didn't know what to think. He rambled. Keats interrupted him and told him he thought that the baby might be the answer he needed. He would now have a child to focus on. "Hey, this is the only thing in your life that I haven't seen you do a hundred per cent," Keats told him. "Whatever you don't know, you'll figure out. You've just got to step up." Steve thought Keats might be right, and got excited, and then panicked, and then he didn't know.

"I remember one day in L.A., and we were just done," Keats says. Wrung out. "There was nothing more to say and we started crying. It was like, 'What's going on? How? Why?' And I knew we couldn't answer our own questions. Finally, I said, 'Today is going to be a better day than yesterday. Yes it is.' And I wrote on his mirror, 'Today is a good day,' because Monty used to write messages on his mirror all the time. And when he saw it, we laughed, and it was like 'Yeah, yeah, okay, here we go.'"

In September, Nick Robinson got married in Greece and Steve was one of his groomsmen. Nick had arranged everything. He lined up boat excursions and dinners for all of the guests. Everyone would get there a week early. There would be twenty-five of them in all, the Peterborough hockey guys—Steve, Keats, and Jay—included. Nick bought all of his attendants "these suits," as he referred to them. "They cost me $600, $700, *together*! They were as cheap as shit. Polyester. I knew they'd destroy them anyway. Dark blue; with white shirts and a tie." Nick, the groom, Smokey Robinson; the others, the Miracles.

Everybody arrived, the boat trips and dinners were great, but Steve went off the deep end. Nick sent him and Keats an email a few nights later suggesting they should ease up a bit. Steve responded right back: "I totally agree, chum. I will clean it up."

"They were both great on our wedding day," Nick recalls, "but that night, Monty was back in full effect. He was firing the DJ up, he had all the tunes going. He was just covered in sweat, and he was a big boy, probably 230, at this point. Just a beast on the dance floor. He had ripped the ass out of his pants, and by the end of the night he was just in his boxer shorts and his dress shirt."

At a different time, this would have been another great "Monty story." But this wasn't funny anymore.

"That was the last time I saw him," Nick said.

———

When Steve returned to California, he checked back into rehab.

Steve's life was filled with people who wanted nothing more than to help him, because they loved him, and because he had helped them. But he never went to them for help.

"When Monty wasn't in a good place and things weren't going well, I wouldn't see him very often," Andy O'Brien says. "He'd just kind of disappear, and only reappear *after* he'd sorted out his problem, when he knew what he had to do and was already doing it." And whenever O'Brien did see him, he sounded good. He had thought things through. He had made a plan, and the plan made sense. So whatever O'Brien or Vally or Keats or Nick, or Steve's parents or brother or sister or former teammates had heard, Steve might have gone through a rough stretch, but he was okay again. That's what they wanted to believe and had every reason to believe, and anything else would mean not believing in him—and out of friendship and loyalty and love, how could they do that? The stories they heard about him were something to wonder about, but not to be concerned about. And Steve came to realize that if he could hide from others, he could hide from himself.

In late September 2014, during his long, quiet hours in rehab, with no alcohol, no workouts, and no CBA to worry about, with nothing to distract him from himself, Steve did a lot of thinking. The past year had been too much: the concussions, the dizziness, the depression, the hope, the loss of hope, the contract buyout, the end of his career, the drinking, the memory problems, the downs, the ups, the downs, the downs, losing himself, trying to find himself. The future. All these balls in the air. Then one more—he was going to be a father—and at Nick's wedding all these balls had come crashing down.

How could he be a father? His own life was out of control. He couldn't even manage himself. He loved kids; he was great with kids. But this would be his own kid, all the time, every day, all of his life. And if he was going to be a father, he would have to jump in with both feet. As Vally said about him, he was a "two-foot jumper," but how could he do that now? But how could he not? He and Chantelle weren't going to get married. They didn't even really know each other. She seemed nice enough, but now she would be the mother of their child. He had come close to marriage a few times, then just when it seemed that some really funny, athletic, beautiful woman was finally the right one and there was no other next step left but to get married, he'd find something wrong with her and be gone. He wanted something perfect for his own family. That seemed worth waiting for. Now a different world was closing in. With women, maybe Steve had a problem with commitment—or maybe he just liked his life the way it was. Maybe he didn't want to give it up. How could he live with himself if he did give it up? How could he live with himself if he didn't? He was stuck.

He had built up his body only to slowly destroy it, then had learned to build it up again. Now he went to the gym only to get respite from the guilt he felt, to stop thinking about the alcohol and the anti-depression, anti-headache, anti-pain, anti-anxiety stuff he was putting into his system. He had loved to learn—he couldn't wait until the adventure of tomorrow; now he didn't have the energy, the attention span, the memory to learn anything. He used to talk to his friends hour after hour, going deep into life and its meaning; now he didn't talk to any of them unless it was to lie to them about all the great stuff he'd been doing and all the great plans he had for the future. He had lived by a code, *The Four Agreements*: Be impeccable with your word; Don't take anything personally;

Don't make assumptions; Always do your best. Principles that had become more than points of pride for him: they had been his identity, that he had now broken again and again.

When he came out of rehab in late September, Steve had made up his mind. He would find stability in his life; he would be a good father to his child; he would move back to Mississauga. So he cleaned up Parros's condo, flew to Toronto, and rented a house less than 200 metres from his childhood home on the cul de sac on Avonbridge Drive; less than a kilometre from the QEW, the highway that would connect him to his child in Buffalo. Instead of his baby being the one additional ball in the air that would send the others crashing down, the baby would be his answer. Keats was right. The baby would be the only ball that mattered; the one that would give meaning to all the others. The baby would give him the reason for the stability he needed, and the will for him to achieve it. The baby would be *the* answer. He bought baby and parenting books. He bought baby furniture. His journals began to change. His thoughts were clearer; his writing more legible.

Then a few days before Halloween, he turned dark again and went back into rehab, only for a tune-up. He was there a week, and then back to Mississauga. Some days were good. Other days were not so good.

In December, he and Missy went to Florida together. They saw the Panthers play; Missy had some clients to see. They went to a concert. They flew back to Chicago for Missy's Christmas party, then drove to Detroit to visit her family. On December 21, Steve drove home to Mississauga. It was his thirty-fifth birthday.

Christmas wasn't so good, but Christmas often wasn't very good. It brought out too many family stresses and complications; and now, too many questions, asked and unasked. But by early in the new year, while things were still never not down and up, he seemed largely fine. Keats drove down from Peterborough and they had dinner. It was the first time Keats had seen him since he got back from Hermosa Beach. "It was like one of the old dinners we used to have," Keats recalls. "It was great. I thought he was getting out of his funk. At dinner, he talked about reading those baby books, and he wasn't drinking, and I was like, 'Okay, all right, we're getting back on track here. This is good.'"

A few days later, in mid-January, Steve chartered a plane and flew to Ibiza, a notorious party island in the Mediterranean off the east coast of Spain. January is not the best month to go to Ibiza, and commercial flights connect the island easily to North American cities. Steve didn't need to charter a plane, but that wasn't the point. He wanted to get away. A few days later, he wanted to get away again and flew to Colombia. He had a buddy in California who knew a shaman who performed religious ceremonies in the Colombian rainforest. The ceremonies involved ayahuasca, a tea-like hallucinogenic brew that indigenous peoples in Peru believe produces spiritual understanding about the nature of the universe and one's true purpose on earth. Ingesting ayahuasca often causes vomiting and diarrhea, which believers see as the purging of negative energy and emotions from one's body that have built up over a lifetime. Ayahuasca can also be potentially dangerous for those with a pre-existing heart condition, and those who take antidepressants.

When Steve got back to Toronto, he called Nick, all excited. "He was explaining all this to me about all these people in this hut in the woods of Colombia, and everyone is throwing up their demons. And he says to me, 'I had a spiritual experience; it's changed my life. Something happened to me there. I am going to get back into the program. I am going to stop drinking. I want to prove to you and to everyone that I can get my shit together. This time it is going to be different.'"

"I will never forget," says Nick. "We talked about how we all had the ability to change the world through even very small actions. He believed that something as little as spending forty-five seconds with the man in the toll booth, or helping a woman carry her baby carriage down a flight of stairs in the subway, or seeing someone who looks dishevelled on the street and making eye contact with him and saying, 'Hi, how are you?' could change the course of that person's day, and, effectively, their life, and therefore the world.

"He really believed that. He thought he could spread something positive, and that it was infectious. He was talking about all this. He was rambling a bit, but not much. He was like the old Monty. When I hung up I remember thinking, 'This guy is going to get his shit together, and I'm going to have this friend back, and our kids are going to be about the same age and they're going to grow up together.'" Nick and his wife were about to have a baby, too. "He just sounded unbelievable, and I remember saying to my wife, 'Wow, this guy is going to figure it out.'

"The next call I got was that he had died."

Steve went to lunch with his dad the day after he talked to Nick. "Conversations with any of my kids tend to happen in snippets because they are all incredibly busy," Paul says. But on this day, neither Paul nor Steve had any other place to go. "We talked about concussions and the state of his brain, and how he was doing, and how he was dealing with that. I asked him what he was thinking of doing next, and right away he said, 'I think I can have a positive impact on people's lives.'" He had said the same thing to Paul six months earlier, when he and his dad had their last long talk.

"It was very clear to him what he was going to do," Paul says. "It was just a matter of figuring out the 'how.' But it was pretty clear he wasn't going to be able to do any of these things unless he was able to clear his mind and get it functioning in a way that he could be reliable and consistent and steady, and not make irresponsible decisions and do irresponsible things, and be so forgetful."

It was Steve's memory loss that most disturbed Paul. He had seen glimpses of it, and heard some stories, but even he didn't know how bad things were. The service manager at Steve's car dealership in Toronto later told him that Steve used to call him every week or two from California. "He was always forgetting where his car was, he said, or he'd lose his keys. He was on Steve's speed dial, so Steve would call and they would locate the vehicle through GPS." Paul was also aware of Steve's charters to Hartford and Ibiza, and his trips to Colombia and to Las Vegas, where in recent months at Caesars Palace he had lost more than $200,000 at blackjack. Steve had never been a big gambler.

Paul and Steve had a complicated relationship. They were alike in many ways: curious, highly independent, full of energy. Both wanted to know everything; both learned fast and wanted to move on. But Paul was more of a manager. He liked to see the big picture, all the problems and possibilities, then encourage those around him to see things the same way as he did, as if they had reached the same conclusions themselves. By contrast, Steve was a doer. He saw what he saw, then jumped into the action. He saw answers first; Paul saw problems. When other hockey parents had imagined their kids as future NHL stars, Paul hadn't. When he had encouraged Steve to pursue other parts of himself, particularly at school, Steve had often taken this as disapproval—as Paul, a baseball guy, not caring about hockey, and what Paul thought mattered to Steve.

A few years before, when Steve was with the Ducks, he and Paul had had one of their talks. Steve asked his dad, "Are you proud of me for making the NHL?" Paul thought for a moment; he knew right away how important the question was to Steve, and how important his answer would be. "And I said, 'You know Steve, I am so proud of you for all the great friends you have chosen. I am proud of you for the people you help and the things that you do. I'm proud that you take good sportsmanship seriously, and I'm happy for you that you made the NHL. I'm proud of you, and I'm incredibly proud of you, for all the hard work you put in because you were never the star of any team, but you made a commitment, and set a goal. And you worked.'

"And I said to him, 'I'm happy, I'm incredibly happy, that you achieved your dream, and I'm proud of you for the man that you've become.' And Steve had this quizzical look on his face, and I said, 'You know, I would be just as proud of you if your talent had only taken you to the AHL, or the ECHL, or if you had worked just as hard at something else and were just the man you are. Those are the things that make me proud as a father.'" Paul pauses as he remembers that day. "I think that satisfied him, although, you know, I think he wanted me to say I was proud of him for being an NHL player."

Steve made several phone calls during the few days after his conversation with Nick and his lunch with Paul. To the usual people: Paul and Donna, Chris, Lindsay, Missy. He drove to Buffalo to see Chantelle and to put the crib together; the baby was due in less than two weeks.

"I noticed that he was talking fast," Missy says, "but when he's excited about something that's his normal." Steve called Daniel Carcillo, too. Carcillo had heard about Ibiza and Colombia and knew that wasn't good, but they talked about other things. Steve told him that he had been to Buffalo to put together the crib.

Steve also spoke to Marty Gélinas. "I had heard some rumblings that he was struggling a little," Gélinas says, so he called him. "Monty said, 'I'm doing great,' and he told me what he was doing and what he was going to do. It sounded like he had it all figured out." Steve called Craig Button, who had been his general manager in Saint John and Calgary. It was Button who had seen something in Steve that no one else had, and signed him as a free agent. He and Steve had bumped into each other a few times in the years since, and in January, Steve called him. A month later, Steve called him again. A ten-minute talk turned into an hour, as was often the case with Steve. Steve told him about being back in Mississauga and about the baby, and about putting down roots. They talked about broadcasting, coaching, and Steve's work with the NHLPA, especially during the CBA negotiations. "It was as if Monty had come to this very busy intersection," Button says, "with all these things going in all these different directions, and he didn't know which one to take." [I said to him,] 'You have a lot to offer,' and yet I was vague." Button seems upset at himself now. "I didn't give him enough of 'You should try this, or that.' But he sounded good, no different from any other time that we talked."

Steve texted his former Saint John coach, Jim Playfair. Playfair hadn't heard from Steve in some time. "It wasn't many words," Playfair recalls, "it was something like, 'The world is awesome. I'm going to be a dad. How is your wife? How are the boys?' Honest to God, there could not have been a more uplifting, encouraging sign."

Andy O'Brien last saw Steve five days before he died. Steve dropped by O'Brien's gym and came to the office to talk with him, but he was busy with another client, so Steve hung around a bit and spoke with some of the CFL players who were training there. Other years Steve would be there every day, always at the same time, and O'Brien could plan his schedule to fit in one of the deep, nourishing talks they liked to have. But now it was over a year since Steve had played his last game (in the KHL). He still came to the gym, but only a few days a week, at different times, for a few minutes or a few hours depending on how he felt and what else he had going on. He and O'Brien didn't talk much that day, but he knew that Steve would be back in a day or two and they would talk then.

Yet it was clear to them that things were not the same. Once, their relationship had been underpinned by mutual expectation, and a little bit of hope. Now the relationship had hope, but only a little bit of expectation. That didn't feel right to either of them, but neither of them knew what to do about it.

"Probably the biggest thing I noticed about Monty during all those good years was the peacefulness he conveyed," O'Brien recalls. "He seemed very grounded. When I first got to know him, he wasn't." That was after Steve had been traded to Florida from Calgary, and before he went into rehab for the first time. "He looked like a guy who was struggling. I used to notice that he would go off on his own at times, into a room, to reaffirm some things through some positive self-talk. And it was really clear that, while he was struggling, he was winning the struggle. He had turned himself into somebody who was very, very much at peace with himself. He was very comfortable in his own skin. He was a very, very zen, calm, confident, secure person."

But "there was this very unsettled feeling about him when he wasn't in a good place." Now, when he came to the gym, "It was like seeing a different person," O'Brien recalls. "It was his demeanour. His ability to maintain eye contact. His ability to just sit still and breathe and not feel the urgency to talk or the urgency to move around." He couldn't do those things. "The quality that had made Monty so distinctively himself was changing. He seemed always rushed. He was on his phone a lot. He just didn't seem present when he was there. He would say, 'Let's go grab lunch,' then not show up, and then show up randomly later on. We would get into a conversation, and midway through it would end because he had something else he had to do. A couple of times he lost his keys; a couple of times in the same week. He lost his passport." O'Brien thinks back. "There were very, very few circumstances in that last year or two where I saw him getting better. He just seemed to get a little worse every time."

For Steve, everything was moving faster. And the faster things moved, the faster he talked, and the more persuasive he tried to be about how well he was doing, the less persuasive he was. He was Monty trying to sound like Monty, and yet he didn't sound like Monty at all. But the spirit was there, they all saw it—Keats, Nick, Button, Playfair—and if the spirit was there, the body would follow. It didn't matter that nobody around him—not his family, not his friends— had grabbed him by the shoulders and shaken him, and stared into his eyes until he couldn't do anything but stare back; for him to see their fear; for him to see the need to do something. Because they didn't feel they had to. He was Monty.

The thing is, though, that Steve would never let anyone get close enough to grab him by the shoulders and pin him down. Especially not his family and friends. He found ways to keep them away. It wasn't hard, just a few out-of-the-blue texts would make them feel close enough that no one would wonder enough to get close enough to see what he didn't want them to see. He was up against it, but he didn't know, and his family and friends didn't know, just how up against it he really was. All of the conditions he was dealing with: the depression, the anxiety, the loss of control, the inability to sleep, the memory loss, the inability to focus, to think rationally, to work things through; the residual pain in his neck, his shoulders, his back and knees, from thousands of little abuses. And all of the drugs he was taking to deal with all of these conditions.

Nobody knows the exact events of Steve's last few days. Nobody really wants to know. Steve was found dead at about 1:30 a.m. on Sunday, February 15, 2015, in the house he had rented only a few hundred metres from his childhood home in Mississauga. He was found by a woman whom he knew a little and who had been with him the previous few days. His autopsy showed evidence of opioids, THC, benzodiazepine, and cocaine in his system. Opioids are used in the treatment of pain. THC is the main psychoactive ingredient in cannabis. Benzodiazepine, or benzo, best-known under the brand name Valium, is used to treat anxiety and insomnia. Cocaine, among other things, can offer a quick "king of the world" rush of energy and euphoria. Opioids and benzo together, or cocaine and opioids together, can result in respiratory depression and death. Steve was thirty-five.

Four days later, Morrison Montador Robidoux was born.

CHAPTER TWENTY-TWO

It was early Sunday morning. Texts, voicemails, emails shot out in a rush, and in them a few urgent words, "It's about Monty. Call me." From Paul and Chris, from Keats, Vally, and Nick, to the rest of the Steve-network. Those who received them knew before they called back, and knew it couldn't be, and knew it was. And when they did call back and were told, many of them cried, went into a deep silence, and made plans to get to Toronto, to do something, to be there.

The hockey world came to his funeral. Teammates from his Lorne Park and Mississauga teams, from the Marlies, Senators, and St. Mike's; from North Bay, Erie, and Peterborough; from all of the NHL teams he played on. His old coaches. His former opponents. People who Paul and Donna had come to know; others they had heard about; many more they didn't know at all. People from other parts of Steve's life, from who knows where, who had nothing to do with hockey.

They were wrecked. Filled with *whys* and *hows*, but mostly they were just sad. At the visitations, a video of moments from his life played over and over, and people stopped and watched it, over and over. Photos of Steve as a child that the friends from later in his life had never seen; photos of Steve as the adult that his childhood friends didn't know. But all of the images, fitting together, so recognizably Steve, culminating with a clip of *that goal.* That face. That dance of celebration.

At the funeral, so many people spoke so movingly, so beautifully. There were shudders and sobs, but there was laughter, too. Steve was just so funny. He was quirky and weird. His life was not to be believed, even for those who had lived it alongside him. Vally told "Monty stories." He made cue cards for himself, a single word on each one. Then, as he told everyone there, in honour of the way Steve talked and lived his life, he shuffled the cards and spoke about Steve in the order the cards appeared. "Which is no order at all," he said. A Monty order. Vally had people laughing until they cried; crying until they laughed. He had them reliving Steve; Steve filled the room.

It was what his family and friends needed. Steve *was* funny. He wouldn't want people sitting around sad. "Hey chum, gotta dance"—that's what everyone told themselves to feel how they needed to feel. It was all too much for Jim Donaldson. Now in his late sixties, Donaldson had been Steve's coach in minor peewee with the Mississauga Braves. Steve was the second of "his boys" he had lost. Donaldson tried hard not to feel the way he did, but to him this was all too sad to be this funny. And he left.

Nick was Nick, serious and thoughtful. He had decided not to write anything down. He wanted, instead, to "speak from the heart." He told one or two stories about times they'd spent together, but most of all he wanted to talk about the impact Steve had had on other people. How Steve believed that a smile, a tiny gesture—even a deep, hard look into a stranger's eyes to remind the stranger that someone else knows they exist—might change a life.

Paul held himself together. When he spoke, trying to find comfort for himself, his words helped others find comfort. "Steve packed seventy years into his thirty-five years of living," he said. To this day, those present at the funeral repeat Paul's words, seeking consolation when they think of Steve.

Months later, all of them were still thinking about Steve's passing. The loss of a son, a brother, an uncle, a teammate, a friend, a father-to-be. The loss of all that energy. His laugh. The loss of someone who made life bigger, richer; who made life feel better. The loss to them, but the loss to Steve. He'd had fifty years of his life ahead of him. Fifty years of who knows what—and that would be the best part—because he was Monty. He lived seventy years in his thirty-five years, and he would have lived a hundred more in his next fifty. That's the lousy part.

"So much attention has been given to the way Monty's career finished," O'Brien said a few months after Steve's death. "About the injury he had, whether it was about CTE, or the concussions; the tragedy of his early death and the tragedy of his life. But to me these are all very, very insignificant stories when I think of Monty. The bigger story was the type of person he was. The way he cared for other people. The phenomenal example he was for others. The value of educating yourself, the value of giving, the value of friendships, and the value of maintaining humility in spite of all that is going on around you. The ability to embrace your teammates and to really enjoy the day-to-day grind at the rink.

"Monty had learned all this. He mastered it. So to me, in spite of everything else that happened, this is what Monty was about. He was just this really, really incredible person. Whether he died at thirty-five or ninety-five, his true story wasn't the tragedy. It was the way he touched people. It was the type of person he was."

At Vail that summer, the winner of the week-long competition between Team Crosby and Team Tavares was presented with the Montador Cup.

CHAPTER TWENTY-THREE

So where do we go from here?

Hockey is a really good game. It has never been played and watched by more people in more different countries than it is now, by more boys and men, by more girls and women, by more people of more different racial and ethnic backgrounds. It has never been played better. Its players have never been so skilled and fit. The NHL has never been a better league. It is where the world's best players play. It is what kids in Moscow and Mississauga, Minneapolis and Malmö dream of. It is the world's league, and the Stanley Cup is the world's hockey trophy.

It is now a game that demands more. More time, more money, more of the life of its players and their families, because when more becomes possible—more tournaments, more special clinics, more months of the year—more becomes the norm. To put it simply, it took Connor McDavid more to make it to the top than it took Sidney Crosby, and far more than it took Wayne Gretzky, Bobby Orr, and Gordie Howe.

It is the same for kids and their families in every field—music, dance, nanotechnology, it doesn't matter what—just as it will be for *their* kids and *their* families in the future. Sports still teach what sports have always taught—teamwork, discipline, hard work, resilience, goal-setting, and goal-achieving—but with the greater commitment now required, these lessons are more intense, and the talent unearthed is even more remarkable. The more you put in, the more you get out. And for those few who do make it, there is recognition and money that might last a lifetime.

For some, sports have other impacts that also last a lifetime—and that is the big new question sports face. For former athletes, bad knees and bad hips might keep you from stretching a single into a double when you're playing with your kids, or they might need to be replaced later in life, but they are no great burden. But what about a bad brain? Not even a brain like Steve's, with CTE, but one that is depressed and anxious, that can't remember, can't decide, can't sleep, can't enjoy; with fifty years ahead of it, that can't live a full life. That *is* a big burden.

Is this now the trade-off for a life in hockey or football? For many, it is. These athletes have made their choice, some will argue. They know what they are getting themselves into. Maybe not when they were kids, but their parents knew, or should have known. And later, after they have lived through the battleground of junior and signed their first pro contract, they're adults and old enough to make their own decisions. Hockey is a tough game. Things happen. Did they think that all the masses of love and money that rain down on them from the heavens are just for donning the jersey and scoring a few goals? This is their danger pay. So some have paid the price, and it is awful and horrible, and something we all wish had never happened, but it has happened, because in a collision sport, it can happen. In life, it can happen.

Do you think Steve knew all this? He knew about the bumps and scrapes. He loved them. He knew about the cuts and lost teeth, the separations and tears. He caused as many of them in others as he ever received himself. He knew about broken zygomatic arches and concussions. He knew about the sort of pain that fills your body and mind and takes over your life. Do you think he also knew about memory loss and suicidal thoughts? Do you think he knew about the sort of pain that fills your body and mind and takes over your life and *doesn't go away*? Do you think he knew about a life that might end fifty years before its time? Do you think he knew that these were part of the trade-off he was making? Do you think Keith Primeau knew about the seven years of pretending to his kids that he was just regular old dad when he was really someone else? What about Marc Savard?

But ask them, ask any player, if they would do it all again *if they knew*. Their answer is: "Of course. I loved to play. I'm a player. I loved playing even before I got all that money and attention. I loved being with the guys. I loved what we all went through together. I love the memories and feelings I still have. Any other life wouldn't be me. Even if I do have some regrets and wishes—and who doesn't—it's who I am. Why would I reject my own life? How could I? If I did, what would I have left?"

But is this the only choice a player has? Was it for Steve? Is it the only choice for all those who now play and who will play in the future? Why is the only choice *this* game—the game as it is now—or not playing hockey at all? Why can't the choice be between a tough, collision game—of cuts and bruises, tears, separations, and breaks, knee replacements and hip replacements-to-be; of rare, random, accidental concussions—and a game that is just as exciting to play and watch, but where a player is less vulnerable to the big injuries, to the injuries that change a life? Why isn't *that* the choice?

What would it take?

Not that much.

Speed is hockey's most defining element. It is also hockey's greatest vulnerability. When a game starts in motion, when the puck moves from stick to stick and end to end, when the baton gets passed from player to player to player to player every forty seconds, when minutes pass—it's as if, as fans, we forget to breathe. Then a whistle, then a breath, then all the *oohs* and *ahhs* that have been stored up. Then the game starts up again. Baseball is a game to savour. Hockey is a game to *feel*.

Almost anybody can do almost anything if they are given enough time to do it. But then try to do the same thing at thirty-five kilometres an hour. Try doing it with a 225-pound (plus equipment) opponent or two coming at you, also at thirty-five kilometres an hour. Speed separates the gifted from the adept, the courageous from the big and strong. From its beginnings, hockey was called "the fastest game on Earth." And it was—and over time it has learned to become faster. The ice got better, skates got better, coaches got smarter, shifts got shorter, players got fitter, rules changed. Now, more than 140 years after that first game in Montreal, there is almost nothing that stands in the way of speed, or the consequences of speed: the increased number of collisions; the force of these collisions; the increased severity of brain injuries.

Sports has always been a compromise between performance and safety. In skiing, we learn to control our speed so that when we go down a hill we don't smash into a tree; and knowing that we won't, we can learn to ski faster. In hockey, we learn to stop so we don't crash into the end boards; so we can learn to skate faster. Safety doesn't need to hold us back. It can make us better. Just as goalie masks made goalies better, safety can *enhance* performance.

Compromise doesn't need to be a bad word. We say that to be good at anything today you have to go all out. You cannot hold back. You cannot leave anything in the tank, because if you do—in business, politics, sports—the other guy won't, and the other guy will win, and you won't. Steve Montador was an all-out guy; if he wasn't, he would never have made the NHL. The league is filled with hundreds of Steves, and family rooms and bars are filled with thousands of guys who were like Steve but who weren't all-out, who did hold back and didn't make it. This is an all-out era. Compromisers lose.

But sport is a contest. It is something *we* create, and if we don't find it interesting or fun, we don't do it, or we change it, just as early pond hockey players did in Nova Scotia in the 1800s in the trial-and-error games they played, and just as it was at McGill in 1875. Just as it has been in all the decades since. Players play and coaches coach, and in all the other non-game hours of their day, they dream and scheme. *What can I do different than I did yesterday? What can I do better? What is more challenging, more interesting, more satisfying, more fun? What can I do that the other guy can't? What can I do to win?* Hockey is always changing and has always changed, because players and coaches have always changed it. Imagine playing or watching a game with no substitutions, no changing on the fly, no forward passes, and five-minute or two-minute shifts played at a snail's crawl.

There is speed that makes sense, and speed that doesn't. We could bring back hooking and holding to slow down the game, but that would reward what we want to reject. We could bring back the red line for offside passes, but that would clog up the open-ice play we want to encourage. We could limit shift length, or changing on the fly, but that would be too restrictive and complicated. We could play four-on-four during regulation time as we once did in over-time, to encourage puck possession, to give extra space to players who have the talent to use it, who know how to slow play down and to speed it up. Modern players have the open-ice skills to play four-on-four without embarrassing themselves. Perhaps someday that might be the answer, but not now.

Two small changes might make a big difference.

First, we need to see "finishing your check" as what it is: interference. *Finishing your check* once didn't exist even as a phrase in hockey, because the circumstances needed for it to occur in a game didn't exist. The game was too slow; finishing your check requires speed. And when it did come about, it happened by accident. In the 1950s and 1960s, passing increased, but not all passes—especially by those more used to advancing the puck by stickhandling—hit their intended target. As a bad pass slid uncontrolled into an opponent's zone, the attacking team did what was left for it to do: it chased after the puck, trying to regain possession. It forechecked. Forechecking was an attempt by a team to neutralize the disadvantage it had created for itself. It was a default strategy, and finishing your check was its unintended consequence.

Forechecking grew in use, especially after the NHL doubled the number of its teams in 1967 and minor league players with minor-league skills became instant major-leaguers. For these players and their coaches, forechecking needed to become not just *a* strategy, but *the* strategy. But for it to work, lesser-skilled players had to get themselves on top of their quicker, more skilled opponents before they could pass the puck to their teammates. If they didn't get there in time, but almost did, and if they were moving too fast to peel off to try and catch up to the play, what could they do? Many just continued forward, crashed into their opponent, and waited for the referee to deliver his justice. But most often—happily, surprisingly for them—that justice never came. The referee let play go on. It was the referees, not the players, who adapted. Their reasoning seemed to be that the player *almost* got there in time; he was going fast and, like a driver putting on his brakes momentarily before powering through an amber light, looked like he was trying to stop. Besides, in an expanded league where disadvantage was the norm, forechecking and finishing your check offered an opportunity for lesser teams and lesser players to balance the larger injustice of an uncompetitive league. Finishing your check seemed a small price to pay.

But what about the injustice to the player who races back to get the puck; who, under the threat of a speeding body coming at him far more prepared for impact than he is, endangers his health or, risking a turnover, makes a pass before he is ready; who, if he does make a successful play, finds that it is to no advantage to himself or to his team because he has been taken out of the play by his opponent anyway—an opponent who, arriving late, made an unsuccessful play yet was not penalized for doing so because finishing your check is part of forechecking, and forechecking is part of the game?

By the mid-1980s, things had gotten even worse. With Gretzky and the Oilers and the growing European influence, with passes and not just a dump-and-chase, the game speeded up more. The most dangerous position on the ice became that of a defenceman, not a goaltender. By far. Goalies wore new cage-style masks that offered nearly perfect protection; defencemen, like Steve, with forecheckers able to finish their checks, played as if they had a target on their backs. Hooking, holding, "obstruction"—almost anything short of tackling was allowed for a time, because referees knew it had to be allowed for defencemen to survive. It was the compromise between performance and safety. But in 2005, after the lockout year, obstruction was no longer allowed. The game got faster, the shifts got shorter, the collisions more frequent and more damaging. And here we are. The forever compromise has been compromised. Safety has lost the day.

But what if "checking" were to become again what it had always been? One player has the puck, another wants it and uses his stick or his body to get it. He *checks* him. If a player, any player, doesn't have the puck, he can't be checked—whether the puck is two feet away from him or sixty—because that is interference. No big deal.

That's how hockey was always played. And the forechecker can go as fast as he wants towards the puck carrier, *he can fly*, but he has to get to him before his opponent makes his pass. If he does, he delivers his hit and gives himself and his team a better chance of gaining possession of the puck. He has earned the advantage he created. If his opponent makes the pass first, his opponent has earned the advantage. That's fair. And if the forechecker arrives after the pass is made and makes the hit, he gets a penalty. This is considered justice even in football: a pass rusher hitting a quarterback after he releases the ball is penalized. And so the forechecker must use his own judgment: *Can I get there in time, or not?* If he thinks *yes*, he keeps going. If *no*, he makes a different play. If *maybe*, he takes the chance or he doesn't. He can go faster or he can go slower, the decision is in his hands. It's up to him to control his own speed—it's not up to his opponent's defence partner, by hooking and holding, to do that for him. The consequences, good or bad, are his to face, just as they should be. The result: controlled speed, not uncontrolled speed.

Speed that makes sense, not speed that doesn't. Advantages that are earned, not advantages that aren't.

No big deal. But a big impact.

The second change is no less rooted in the game: no hits to the head. Players and officials since the first game at McGill have recognized the vulnerability of a player's head. A "high stick" isn't something that is delivered to a shoulder, or an "elbow" to a hip. Rule-makers created high-sticking and elbowing penalties specifically to protect a player's head. Yet there was an exception to this. For the head to get this special treatment, it had to be *up* at the moment it was struck. If it were *down*, the hit was considered to be the player's own fault; the player was "fair game."

This thinking arose out of the pre–forward pass era. If a player has to advance the puck up the ice without passing, he has to stickhandle, which is easier to do if he looks down at the puck. If stickhandling is to his advantage, the checker should have some advantage of his own. So a hit to the head by a stick or an elbow wasn't allowed, but a hit to the head by a shoulder or a hip delivered with far greater force was not only okay, it was glorified, as long as the player's head was down. It was a "freebie."

There is another category of hit to the head that is also penalized but accepted: a fist to the head. It's because it has never been seen as a hit to the head, but rather as fighting, which isn't possible without some fists to the head, and is acceptable, in fact essential, because hockey is a passionate game different from other passionate games, and an emotional release is necessary.

This is all very admirably consistent, but even if it wasn't, even if you wanted to make these changes, you can't. The game is the game and you can't change the game. It has been played this way for a hundred years. The traditionalist "knows" this.

But it hasn't. The game has changed constantly. It is *always* changing. The traditionalist is blind. The game is much faster. A hit to the head from a shoulder moving at thirty-five kilometres an hour is a hit from a much different shoulder than fifty years ago. A hit to the head from the fist of a bigger, better-trained fighter is a hit from a much different fist. A hit to the head today is a much different hit.

The NHL created a Department of Player Safety in 2011 to review dangerous hits and to impose additional penalties if the department believes they are warranted. The length of suspension for these infractions has been increased over time. Rules have been changed. When it was implemented in 2010–11, Rule 48.1—"Illegal Check to the Head," as the league terms it—stated that a "lateral or blindside hit to an opponent where the head is targeted and/or the principle [sic] point of contact is not permitted." Which, put a different way, meant that a check was "legal" if the player who was hit ought to have been able to see the other player coming, if his head wasn't targeted and/or made the principal point of contact by that player, even if the result were a serious and injurious blow to the head. Rule 48.1, as amended, now penalizes "a hit resulting in contact with an opponent's head where the head was the main point of contact and such contact to the head was avoidable." It also adds this as clarification:

> In determining whether contact with an opponent's head was avoidable, the circumstances of the hit including the following shall be considered:
> (i) Whether the player attempted to hit squarely through the opponent's body and the head was not "picked" as a result of poor timing, poor angle of

approach, or unnecessary extension of the body
upward or outward.

(ii) Whether the opponent put himself in a vulnerable
position by assuming a posture that made head contact on an otherwise full body check unavoidable.

(iii) Whether the opponent materially changed the
position of his body or head immediately prior to
or simultaneously with the hit in a way that
significantly contributed to the head contact.

In other words, for a hit to be illegal, the head no longer needs to be
"targeted" but must not be "picked", and no penalty will be given if
the player has his head down, or if by changing his body position he
attempts to draw a penalty on his opponent. Other hits to the head
remain "legal." Within and beneath this careful language, the
message and its intent are clear: a thunderous hit from a shoulder
that stops an opponent in his tracks *should* be legal even if it also
makes contact with an opponent's head, no matter the injury suffered, because that is hockey; and therefore it is legal. Simply, if it
looks like a hockey hit, it *is* a hockey hit. The message is about the
hit, not the head.

The rule is misguided in another way, too. It reflects a style of
game and an understanding from another time. We don't stickhandle up the ice anymore. We don't give up the puck, to dump it
eighty feet ahead, and fight to get it back. We don't forecheck a lot.
We want possession of the puck. So we pass. And when we pass, we
look up and look around for teammates in open ice. When we see
them, we drop our heads and bodies a little to see the puck on our
stick, and make the pass. We might see an opponent coming right at
us, but our vision is narrowed; we focus on the target of our pass.

When we receive a pass, our heads and bodies also drop a little. We are looking at the puck. We might see an opponent coming at us, but again our vision is narrowed; we focus on receiving the puck. At the very moment we pass the puck or receive the puck, and for a few moments after, we are completely vulnerable. We are bent over slightly, but more significantly our attention is elsewhere. We are in no position to defend ourselves. This is not about a blindside hit, a blow from someone out of our possible range of sight. At this moment, we are blind to almost everyone. And because this is about a pass, the puck receiver, free of carrying the puck, breaking towards open ice, is moving fast. Think of Keith Primeau's stories about how he got injured. Of Marc Savard's. How many players received those injuries in open ice, cutting across the middle, getting hit by what they didn't see? And because they didn't see it coming, they weren't ready for the hit they got.

Think of bighorn rams running at each other, their heads and horns colliding at thirty kilometres an hour. Think of gannets hovering in the air, spotting their target fish below, diving straight down, hitting the water at ninety kilometres an hour. They can do this because of bone and muscle and brain structures that have evolved over eons of time; and because they are ready. The bighorn rams can see each other, the gannets can see the water; they know what's coming, and they prepare. Football and hockey players build up their neck muscles to give their heads a stability that bone, muscle, and brain evolution didn't provide. Players' necks are no longer just narrow bridges that connect body and head, but they splay out wide towards the points of their shoulders. Before the moment of impact, they flex. But when they are not ready for a hit, when players like Matt Cooke appear from nowhere, they can't flex. Imagine a bighorn ram that doesn't see another bighorn ram coming at him, a gannet that doesn't see the water. Why don't helmets work against concussions? Why do most concussions seem to occur with whiplash-like hits that spin a head around, and not from direct, straight-ahead blows? One you see coming, the other you don't. One you have prepared for, the other you haven't. Big football linemen hit each other hard and often; whippet-like wide receivers and defensive backs much less so. Yet the receivers and backs seem much more vulnerable than the linemen. One plays a predictable game; the other doesn't. In today's hockey, in the passing game, speed, distraction, vulnerability, finishing your check, hits to the head—it is a mix made for disaster.

Further within and beneath the language-dance of the NHL rule book is another message: you have to allow some hits to the head because you have to allow fighting—because fighting has always been part of hockey, hockey fans love fights, and fights symbolize the commitment of a hockey player that is different from the commitment of a player in all other sports. A hockey player, literally, is willing to *fight* for his team. Besides, if there were no fights, and no outbreak of stick-swinging resulted; if players were to find their emotional release in skating faster and checking harder, as almost every player now does and has always done; then real hockey fans, in their passionate defence of fighting, would have been living a lie all these years, and fighting's critics, who aren't real deep-in-the-bone fans, would win, and that would be too humiliating to take. So if some hits to the head are allowed—in fighting—to remain admirably consistent, then why not allow other hits to the head under some word-parsing, hair-splitting and re-splitting logic that is able to wrap itself up as part of the essence of the game?

Consider this exchange on Sportsnet between John Shannon, Scott Morrison, and program host Jeff Marek. It happened the day after Pittsburgh Penguins defenceman Kris Letang hit Marcus Johansson of the Washington Capitals during the second round of the 2016 playoffs. Johansson had carried the puck through the neutral zone over the Penguins' blue line, and was bent slightly forward trying to fend off a check by Evgeni Malkin. Johansson's head was up, his eyes were on his teammate Justin Williams to his left. Letang, unseen, was to his right. Johansson made a one-handed pass to Williams; Letang moved towards Johansson, braced himself, raised up his body to make contact with him, his upward momentum taking his skates off the ice, his left shoulder striking Johansson in the head. Letang had been averaging nearly thirty minutes of ice time a game for the Penguins, almost seven minutes more than any other Pittsburgh player. Later in the day of this interview exchange, the league would decide if Letang was to be suspended.

Marek begins, commenting about Letang's hit: "A little bit late. A little bit to the head. What are we looking at here?" he asks his panel.

"Certainly late," Shannon concurs, "I think it's a single-game suspension."

In the previous game of the series, the situation had been reversed: Brooks Orpik of the Caps had hit Olli Määttä of the Pens in the head with his shoulder after Määttä had made a pass, for which Orpik received a three-game suspension.

"Everyone in Washington will say why not three?" Marek says.

"Because it wasn't as late," Shannon replies. The panellists comment back and forth.

"And the injury factor, too," adds Morrison. Johansson later returned to the game; Määttä did not.

But Shannon adds a caution: "Apparently Johansson is not skating this morning," he says. He and Morrison agree that sometimes concussion symptoms are delayed.

Marek pushes back at them. "Would that be gamesmanship, though?" he asks. With Johansson not being at practice, he *might* be injured, and the NHL, not yet having made public its decision on Letang, might impose a longer suspension if it thought Johansson might not be back in the lineup for the next game. "Not to sound crass," Marek continues, "but let's be honest about it. It's the playoffs."

Shannon still believes the league will decide on the basis of the lateness and the nature of the hit. "I don't think they view it as a blatant headshot," he says.

Marek talks again about "lateness": "As we've seen with the Department of Player Safety, they will allow a player to 'finish the check,'" he says. "And listen, Kris Letang, you see a guy who has his head down coming over the middle, a chance to get a lick in on him. I get it."

Shannon picks up on Marek's point: "And the question becomes, did he have enough time to swerve away and miss Johansson? And at the same time, where was Johansson's head position . . . was he reaching? Was he down?"

Morrison has a question: "Do you think [Letang] thought, 'I'm going to rock his head,' or, 'I'm just going to rock him?'"

"I think he was just going to rock him," Marek says. "He said, 'Oh, oh, here comes a freebie. Tomato over the blue line and I'm going to get a shot in.'"

Morrison and Shannon agree again that Letang will get a one-game suspension, then suddenly realize the implications of their own judgment. "It's devastating," Marek says. "Letang has been outstanding...." They make no mention of the impact on Johansson.

Listen to this other exchange from a Monty Python skit between a customer, Mr. Praline, and the owner of a pet shop. Half an hour earlier, Mr. Praline (John Cleese) bought a parrot from the shop. He's back to register a complaint. "It's dead," Mr. Praline says. The shop owner (Michael Palin) says it's not dead, "It's resting," and goes on to say, "Remarkable bird, the Norwegian Blue. Beautiful plumage." Mr. Praline takes the bird from its cage, whacks its head on the counter, throws it up in the air, and watches it hit the floor. "Now that's what I call a dead parrot," he says. "No, no, it's stunned," the shop owner replies, then goes on to claim, "Norwegian Blues stun easily." Furthermore, it's not dead, "It's probably pining for the fjords." Mr. Praline is getting more and more exasperated: "It's not pining, it's passed on! It's expired and gone to meet its maker!" To Mr. Praline, this isn't about a parrot that is resting, stunned, or pining. It's not about it being a Norwegian Blue with beautiful plumage. This is about a dead parrot.

For hockey, this isn't about a player having his head down, an opponent with an intention to "rock him" or "rock his head," or getting a "freebie," or the importance of Kris Letang. This is about hits to the head. It's not about the perpetrator, it's about the victim. It's not about the perpetrator's inconvenience/pocketbook, it's about the victim's brain/life/future.

For hockey and its future, this isn't about winning or losing old arguments. It is about tomorrow. Focus here. Start here. And when the debate gets quickly ensnarled—"but *what if* one player is shorter, *what if* one is leaning forward, *what if*..."—don't be distracted from the point. This is about hits to the head. Look at the NHL's own rule book for guidance and precedents. Rule 60.2 imposes a minor penalty for "*Any* contact made by a stick on an opponent above the shoulders," . . . Rule 60.3, a double minor or more is imposed if an injury results from the contact, *whether accidental or careless*" [italics added]. It doesn't say anything about whether the head was the main point of contact, or whether the player should have seen the stick coming, had his head down, or tried to draw the penalty. The league decided that a stick to the face is a bad thing. It is a penalty. Automatic. Period. As Sidney Crosby put it, concerning hits to the head and the league's rules after he received his concussions, "If a guy's got to be responsible for his stick, why shouldn't he be responsible for the rest of his body?"

If some players try to abuse the rule, if they duck into a hit to draw a penalty, then penalize *them*. Use another of the league's rules as a model: if a player attempts to draw a penalty by diving or "embellishing," *he* gets a penalty, and perhaps a fine or a suspension as well. Do the same for a player "embellishing" head hits, except make the penalties more severe because his embellishment isn't just about normal gamesmanship, about getting away with something in order to gain an advantage before the other guy does the same. Diving is merely an embarrassment to the game; head injuries are dangerous—to the player, but also for the game. Drugs, gambling, and performance-enhancing substances all bring a sport into disrepute, and all are treated harshly by the league. But players who end up damaged also bring a sport into disrepute. The NHL can say with complete justification that any attempt to undermine the league's efforts to make the game safer for its players, and for youth players everywhere, for now and for the future, will not be tolerated and will be dealt with severely. A serious problem requires an equally serious response.

If there are other abuses, you find other answers. Players and coaches adapt. They adapt to every next game and to every next moment and to every situation and opponent in a game. That's what they do. And they do the same with rule changes. When a player hits an opponent in the face with his stick, he now skates directly to the penalty box; no argument. If his opponent is cut, he gets a four-minute penalty; no argument. If he shoots the puck over the glass in his own zone, it's a two-minute penalty; no argument. There should be no such thing as a "legal" check to the head. A hit to the head is a penalty. No argument. If a player applies physical force to an official, depending on intent and severity, the player is suspended for either three, or ten, or twenty games. No argument. In that rule, "intent to injure" is defined as "any physical force which a player knew or should have known could reasonably be expected to cause injury." Do head injuries not similarly undermine the integrity and reputation of the game? A hit to the head with the intent to injure is a suspension. No argument.

And if this isn't enough to dramatically reduce brain injuries in hockey, then you do what any serious person does when you face a serious problem. You do more.

This is where Gary Bettman comes in.

We all want the problem of brain injuries in hockey not to be a problem at all. We want it to be about bad luck, or something that comes out of nowhere, is treated, and is gone. We want for it to be about media hype, not reality; about public awareness of something old, not something new. We got through it before, we'll get through it again.

We also want brain injuries to be about fighters and fighting, so that if we create the conditions for a style of game where the specialty fighter is too great a risk to have on the ice, then the problem goes away.

We want the answer to be equipment. The most advanced helmet manufacturers will tell you there is no evidence that better helmets reduce concussions. Then they go on and talk excitedly about the new helmets they are creating, and leave the impression that a better helmet *is* the answer. It isn't. Neither is a better mouth-guard. Or some new whole-body, rotationally stimulating gyroscope. None of these things are the answer. But where there is fear, there is hope. And hope-mongers. And we waste time, we waste careers and lives looking in the wrong direction.

We want for the answer to be science. We want for science to *know*, and to know *now*, and to tell us what to do. But science doesn't work that way. Science is about learning. Whatever we think we know at any moment is only a placeholder for what we will know better in the future. The sun revolved around the earth until we learned differently. Science isn't about certainties; it is about like-lihoods, probabilities, the best we know at any particular moment. But decision-makers want certainties, and decision-makers who don't want to make certain decisions hide behind science. They may acknowledge what we know, but they focus on what we don't. They create doubt. After ignorance, after denial, after counter-argument and strategic inaction, after being put into their final corner, that's what tobacco companies did. That's what lead and asbestos companies did. That's what climate change–deniers do. They will wait for science to know, they say, because after all, we are all serious, modern, evidence-based people, aren't we? Except science takes time, and games are played tomorrow. And people, and players, have to live with the consequences of tomorrow.

We want for awareness to be the answer. We want more scien-tific studies, more articles and books, more movies and investigative reports, more tragic stories of players we have watched and loved. We want to build a mountain of awareness so tall, so sad, so unmis-takable, so unmissable that something will be done because it *has* to be done, because how can it not be done?

But the answer for brain injuries is not awareness. It is not in our scientists, researchers, equipment manufacturers, writers, or film-makers. It is in our decision-makers. Decision-makers decide.

Decision-makers will sometimes make the *wrong* decisions. When we think back a hundred years or more—to slavery, or to the absence of women's rights—we now wonder, "How could they have been so stupid?" When we think back fifty years to the dangers of tobacco, or lead, or asbestos, we wonder the same. Fifty years from now, people will look back at *us* with the same incredulous eye about something. What are we getting wrong? Climate change? Global pandemics?

In sports, it will be brain injuries. *How could they—how could we—be so stupid?*

Gary Bettman has been the commissioner of the NHL since 1993. His contract, extended last year, expires in 2022. It seemed to some that Bettman might choose 2017 as a natural moment to retire. He would be sixty-five. It would be his twenty-fifth year as head of the league. It is the hundredth anniversary of the NHL, a season that will be filled with celebrations. Many of the league's loose ends have been tied up—the CBA has been signed, TV contracts are done, the league's franchises are more valuable and more secure than they have ever been. The expansion process is in order. One new city has now been added, Las Vegas; the other, Quebec, is use-fully in place as a credible threat in case the fans or local politicians of an existing franchise's city get too difficult. But if 2017 seemed a logical time to others for Bettman to leave, it didn't seem logical to Bettman. His first ten years as commissioner were not easy. He is just getting good at this. Why would he want to leave?

Gary Bettman is in control of the hockey world. The league's Board of Governors are his bosses, but since he won his salary cap in

2005, crushing the NHLPA, no one stands in his way. The International Ice Hockey Federation (IIHF), led by René Fasel, has members from seventy-four countries, but it is no rival. Nor is the KHL. Nor is Hockey Canada, or USA Hockey, or any other organization or federation in any other country at any level. They set local priorities. They administer. But it is the NHL that sets the direction for the sport, and Bettman sets the direction for the NHL. Moreover, all these other leagues and organizations *need* the NHL. They need it for hockey to be a showcase Olympic event. They need it for the World Cup, World Championships, and World Junior Championships to generate the money they require that drives their programs and pays their salaries. Those who have power understand power. The rest of the hockey world understands that on any issue, if they disagree with the NHL, it is at their peril. They are all beholden to Bettman. They know it, and Bettman knows it.

The NHLPA, led by Don Fehr, has become more formidable again. Fehr's background is in baseball; he was a protégé of Marvin Miller. Miller had worked for the United Steelworkers union. The Major League Baseball Players Association (MLBPA) had played rag doll to the teams' owners for decades; Miller was hired to head the MLBPA in 1966. With his silver-white hair and carefully ordered moustache, Miller was an elegant street-fighter. By the time he retired in 1983, replaced by Fehr, the players were no longer tied forever to their teams by a reserve clause, they could become free agents—and salaries increased, on average, from less than $20,000 a year to over $300,000. Fehr learned from Miller some basic lessons: Owners are owners; and players, no matter how highly paid, are labour. Owners need to run things, owners insist on that; the best a player can do is to be paid his proper piece of the revenue pie. Owners will never stop trying to control players or to destroy their unions. Players have to be good teammates, whether in a steel plant, or on a playing field or rink, and be ready to fight back. The average salary in baseball increased tenfold during Fehr's tenure. He stepped down in 2009, and a year later was hired to head the NHLPA.

Fehr is an old-style players' union leader. Like Miller, he isn't looking to run the game; he leaves that to Bettman. He wants, instead, like Miller, to maximize the benefits to the players, by increasing the size of their piece and/or the size of the pie. Historically, that has meant mostly economic benefits, and in negotiations Fehr doesn't like to push non-economic, quality-of-life issues too far, knowing that Bettman will ask for economic concessions in return. But for Fehr to do his job now, he needs to know that it is one thing for players to earn as much money as possible, and another for them to be able to enjoy that money for as *long* as possible. For them at age forty-five and seventy-five to have a brain that works. That is the job of a new-style players' union leader. Safety and quality of life were Miller's job when he acted for the steelworkers; less so when he worked for the baseball players. The game is changing for Fehr.

For Bettman, Don Fehr is an exacting adversary, but he is no threat. Bettman today finds himself in a place he must surely have never expected. Almost twenty-five years after being named NHL Commissioner, this American who never played the game is on top of the hockey mountain. He has no peers; he has no rivals. But with this reality has come another, that Bettman must also have never expected and may not want, for ultimate authority brings ultimate responsibility.

Brain injuries in hockey *are* a problem. Indisputably. But in hockey, unlike in football, there are answers—good, doable, real hockey answers. That is the exciting part. The other exciting part is that Gary Bettman is the right person to implement these answers.

He can play "clever lawyer" at times, and he is good at that. With wordplay and an incisive mind, he can get himself out of almost any corner in which he finds himself, an essential skill for a sports commissioner. He can also play pit bull with lawyers representing players on the other side of the class-action concussion case, tell them that their lawsuit is going nowhere—as he has done—because they can't prove causation, because they can't prove that the players' injuries happened in the NHL, and not in junior, or playing minor atom with Lorne Park, or from jumping down the front stairs as a two-year-old, or from abusing alcohol or drugs, or from hitting their head on a doorframe. And he can employ the same adept belligerence when governments attempt to intervene.

On June 23, 2016, U.S. Senator Richard Blumenthal of Connecticut wrote to Bettman, introducing himself as the "Ranking Member of the Senate Subcommittee on Consumer Protection, Product Safety, Insurance, and Data Security," the committee that has jurisdiction over professional sports. Blumenthal noted in his letter the NFL's admission earlier in the year about the connection between playing football and CTE, and described the subsequent NHL response to the question as it relates to hockey as "dismissive and disappointing." He pointed out to Bettman the NHL's obligation to ensure the safety of its players and to "engage in a productive dialogue about the safety of your sport at all levels" because of the NHL's position as "the premier professional hockey league in the world." He also reminded Bettman that because many NHL teams "play in arenas financed in part or in whole by taxpayer funds and [because of] the hundreds of thousands of American children playing hockey, government oversight into the safety of your sport is appropriate, and a matter of public health." At the end of his two-page letter, Blumenthal then set out nine questions for Bettman to answer, most of which related to CTE and its link to hockey; others to fighting, concussion protocols, and the education of players as to the risks of playing hockey. He asked for Bettman's response by July 23.

On July 22, Bettman responded in the form of a twenty-four-page letter, single-spaced, that included forty-one footnotes and ten bibliographical references. He began by suggesting to Blumenthal that some of his questions "appear to be premised on misconceptions that have been repeatedly promoted in the media" by the lawyers for former players now "pursuing concussion-related litigation against the NHL," and that he, Bettman, appreciated the "opportunity to correct the record" and to share with Blumenthal "important information on these topics."

"First and foremost," Bettman wrote, "we want to state in no uncertain terms that the health and safety of NHL players is a top priority for the NHL and its member Clubs, as well as the National Hockey League Players' Association." He continued: "Without question, head injuries, including concussions, are worthy of serious medical attention and care." The rest of his letter continued in the same tone of restrained concern.

He told Blumenthal about the actions the league and the NHLPA have taken, its "Concussion Program"—"the first of its kind in professional sports"—its Injury Analysis Panel and "Concussion Protocol"; its Department of Player Safety, "the first of its kind in professional sports," and the league's efforts to inform the players of the risks of head injuries. But most of his letter focused on CTE and scientific studies, and how they relate or don't relate to hockey and the NHL. He mentioned the work of several brain experts, and most notably cited the 2012 Consensus Statement on

Concussion in Sport, of which many of these experts—including Dr. Robert Cantu, he noted—are signatories, and which states, as Bettman quoted, "It was . . . agreed that a cause and effect relationship has not yet been demonstrated between CTE and concussions or exposure to contact sports."

"It is this medical consensus," Bettman explained to Blumenthal, "that has guided, and will continue to guide, the NHL on the topic of CTE, unless and until there is sound medical evidence to the contrary that can be relied on." Further, Bettman added, no scientific study has shown a causal link between concussions suffered by players in the NHL and degenerative brain diseases of any sort.

Bettman ended with a cautionary tale. It was the story of former NHL player Todd Ewen, a "so-called 'enforcer,'" who committed suicide, who had suffered some of the effects of concussion, including depression, but whose autopsied brain showed no signs of CTE. Bettman quoted Ewen's wife: "[W]e were sure Todd must have had C.T.E.," she told the media, and because symptoms don't always mean the existence of a concrete condition, she said she hoped that anyone suffering from these kinds of effects "takes heart." Bettman concluded:

> This, sadly, is precisely the type of tragedy that can result when plaintiffs' lawyers and their media consultants jump ahead of the medical community and assert, without reliable scientific support, that there is a causal link between concussions and CTE. Certainly, a more measured approach consistent with the medical community consensus would be a safer, more prudent course.
> I hope this letter satisfactorily responds to your inquiry.

Three weeks later, on August 15, Blumenthal replied. It was short, only three pages (and four footnotes) long; more angry than it was substantive. "Your letter notes concussions can cause 'long-term lasting effects' like 'permanent brain injury' and 'permanent brain damage,'" Blumenthal wrote, and yet "the league appears unwilling to consider even the possibility that concussions cause CTE." Later he asked Bettman, "Why is the league so seemingly indifferent to seeking more answers—choosing to sit on the sidelines and 'monitor' the matter instead of urging and supporting research needed to determine whether hockey players are at a heightened risk of debilitating disease?" Before suggesting what Bettman's actions might be instead, Blumenthal again reminded him that "The NHL is a big business that produces an estimated $4 billion in annual revenue. Much of that revenue is made possible by exemptions from federal antitrust laws, and local and state support. The NHL has a duty to behave responsibly in light of its public trust." Blumenthal then wrote, "I urge you to establish a foundation to support independent, impartial research and understanding about the science of head trauma and brain disease later in life, vigorously working to uncover the truth rather than degrading science and the game."

Blumenthal concluded, "I invite you to re-evaluate and resubmit your response, delineating the steps you are taking to advance science instead of dismissing it."

A few days later, Bettman was interviewed by Jonas Siegel of the Canadian Press, and was asked about the senator's second letter and whether he was surprised "how much Senator Blumenthal's attacks . . . have gained traction."

"[What Blumenthal said] hasn't gained any traction," Bettman replied. "In fact, we haven't heard from very many people about it all." Bettman went on to repeat that he believed "medical and scientific decisions should be made by scientists and physicians."

Several months later, Bettman had not accepted Blumenthal's invitation to "resubmit" his response.

The exchange between Bettman and Blumenthal presents two additional potential problems. Blumenthal's suggestion that the league create a foundation to support independent research into head injuries allows Bettman to get off the hook with embarrassing ease, and with little effect on the problem Blumenthal seeks to abate. The NHL can commit a few or many millions of dollars to research, and while this learning happens, more players will have their future lives diminished. Science takes time. "Scientists and physicians" can only offer the best information available at the time. They can only help inform decisions made by decision-makers, who choose whether or not to be informed, and choose whether or not to understand the implications of that best information and whether they wish to apply it. Decision-makers make decisions.

Secondly, with his letter and his comments, Gary Bettman is pushing himself deeper and deeper into the wrong corner, from which he might find it easier to continue acting as his "clever lawyer" self, his least effective self. To fight back doggedly rather than deal with the embarrassment of changing his position—to his own detriment, and to that of the game and its players. Sometimes we lose when we win. Bettman can mock the weakness of the players' position, he can sniff at a senator's pretence of power and dance rings around him, he can delight in the incisiveness of his lawyer's logic and win a lawyer's fight and save his owners some money. Or he can take off his lawyer's hat and do far more than that, because he has far more than that in him.

337

Look at the changes that have occurred during Bettman's time as commissioner. Hockey has become a more international game. NHL players are now born and raised in Slovenia, Denmark, Norway, Austria, Kazakhstan, Hungary, and many other countries besides. That wasn't possible even a decade ago. And hockey has also become a more North American game. There are NHL teams in more winter-free U.S. cities, and local kids are playing there—in California, Texas, Arizona, Florida—and making the NHL. That once wasn't possible either.

Hockey is no longer just a Canadian game played by Canadians who fight and have no teeth. It has become a mainstream sport to more people in more places. Others, in other countries, have the right to feel—and do feel—that hockey is theirs, too. But two things stand in the way of the game's greater possibilities. Cost—equipment, ice time, travel to faraway places for year-round tournaments and games—and, most crucially, brain injuries. More than thirty years ago, Paul Montador couldn't resist the instincts and pleadings of his kids to play hockey. Today, more Paul Montadors and Donna Montadors do.

Again, for Gary Bettman, it doesn't have to be this way.

Football and the NFL offer useful lessons. Football is a much more popular game in the U.S. than hockey. On Friday nights at high schools, on Saturday afternoons on college campuses, on Sunday afternoons in big-city stadiums, football offers a chance for people to gather in the same place at the same time, to feel and express the pride and community connection we all want to feel—about *our* high school, *our* college, *our* city, *our* country, not just about a game or a team. To lose football would be to lose all this. The stakes are high. This is why football needs to find an answer. But football's answer is more difficult than hockey's. Football *requires* body contact. One set of bodies must force the ball up the field; the other set of bodies must resist—and out of their athletic crouches, the players' heads are involved in the action on almost every play. And every year, bigger bodies are harvested, learn to hit harder, and get hit harder back. Those who make decisions in football, with its much more elusive answers, can't run away from the problem fast enough. Their biggest hope is that football is, and can be made to be, too big, too important to fail. But now, because the public won't let them run away entirely, its decision-makers are running away by *appearing* to run towards the problem—by giving money for brain research, by mandating independent doctors to search out head hits from high in a stadium, by restricting body contact in practice, or for kids of younger ages. Between Monday and Thursday, in the stories we read, everything seems hopeful, if not quite promising. Then on Friday night, Saturday and Sunday afternoons, we watch, and we see the unbelievable collisions; we see Roger Goodell's deer-in-the-headlights look of frozen fear whenever the subject of concussions comes up, and it is utterly clear: the NFL has no answers.

Hockey has always been in the shadow of football in the U.S.; the NHL in the shadow of the NFL. Bettman and others in hockey resent that the NFL's dark shadow of brain injuries has spread over the NHL as well. They argue that the NFL's problem is *its* reality, not the NHL's. Yet the NHL has also benefited from this shadow. The football situation is so bad, and the NFL's handling of it by Goodell so inept, that hockey and Bettman have come to look better—and better than they deserve—by comparison. The NHL has before it both obligation and opportunity. If it chooses to do what needs to be done to prevent brain injuries, what it is capable of doing, the *NHL* will be the league, not the NFL, and *Bettman* will be the commissioner, not Goodell, that does things right. The NHL can emerge out of its own brain injury shadow more respected, its reputation enhanced. The actions it takes might also put pressure on the NFL to do much better, an important result for everyone. And the bonus for Bettman and for the NHL, its great good fortune, is that all of this is much easier to pull off in hockey, because hockey's answers are easier.

But first, Gary Bettman and the NHL must recognize that brain injuries are a big problem in hockey. They need truly to believe this, and to reflect it in the ambition of the actions they take. They need to approach this not as a *yes, but*... problem: "Yes, but... the NFL's problem is worse"; "Yes, but... we were the first league to establish an injury committee"; "Yes, but ... the scientists don't really know yet." *Yes, but*... is the wrong distracting message, and it gets us nowhere.

Bettman and the league need to recognize, too, that this is not a CTE-or-nothing problem; this is about diminished careers and diminished lives. In the First World War, soldiers who emerged from the conflict and looked fine but were somehow different were said to have suffered "shell shock"; in the Second World War it was "combat fatigue"; now it is PTSD. But they are the same thing. Many of these soldiers couldn't cope; many became alcoholics. They were seen as weak, as cowards, unable to move on to the responsibilities of the rest of their lives. Now we know their conditions were real, that their brains were physically damaged by the concussions that came from the shock waves of exploding devices around them. Steve Montador looked fine. Keith Primeau, Marc Savard, and scores of other former NHL players look fine. But their injuries are *real*. Bettman and the NHL need to know this.

They also need to recognize that the game has changed. Gary

Bettman has never played the game, but the hockey guys he has around him have never played *this* game. That is the crucial point. The game that, in their memories, moved a hundred miles an hour when *they* played, didn't. The crushing hits they felt and gave that were like no others *ever*, weren't. The "bell rung" they experienced is not the "bell rung" that today's players are experiencing. These are dedicated hockey guys, but their "inside the Beltway" thinking has put the sport and its players at risk. Trust what you see, not what you know. This is a faster, more exciting, more dangerous game. The evidence is found not in our memories, it's in the injuries we see around us. And the message of these injuries is clear: A hit to the head is a bad thing. Hard hits, frequent hits to the head, are very bad things. Head hits rattle the brain. They change the brain and change the person. Life becomes different. Steve Montador, as his dad said, lived seventy years in his thirty-five-year life. But in the last year, except at brief moments, Steve wasn't Steve anymore. He was gone. The worst thing, what is most haunting and tragic about Steve, is that when he changed, when he died, he thought it was his fault.

Gary Bettman and the league also need to understand that the players have *not* changed. Players play. They will play no matter what doctors or their wives or their families tell them. And hockey players will do all this, and *more*, because they are hockey players. A lawyer involved in class action concussion suits in football and hockey was asked why so many more football players than hockey players sue. He said that when football players leave football, they hate the game and hate their team. When hockey players leave, they love the game and love their team.

But a player who loves to play is a vulnerable player. He is one who, if the moment requires that he go through a brick wall, will.

He can be manipulated. He can be seen as weak. He can be treated by decision-makers with contempt, not respect.

Gary Bettman and his lawyers may be right. Perhaps some of the brain damage suffered by the NHL's players happened in junior or on a backyard rink. Perhaps this absence of certainty *will* affect causation and liability, at least to a degree. But if, as the NHL contends, a player does arrive in the league already damaged, if that is a real possibility, perhaps the league might take note of the importance of this, might decide that it doesn't want any pre-existing condition to be aggravated, and implement the rules necessary to protect these players. After all, they are *its* players. After all, as the league likes to say, this is the "NHL family." Whether the league has a legal obligation to do so or not, maybe it would be a good thing—the right thing—to do. And it is so doable.

All of us need to be saved from ourselves at times. Steve did. It's why we have traffic lights. The league needs the players to go full out when it's time to go full out. The players need the league— its doctors, rule-makers, and decision-makers—to say stop when it's time to stop.

The NHL's whole system of doctors, rule-makers, and decision-makers faced a test late in the 2016–17 season. In the endlessness of a regular season, players are in and out of the lineup, injuries happen, every day is a new story, nothing has visibility for long, the season goes on, and life goes on. In the playoffs, each series is its own dramatic saga. Every game is anticipated and argued about. A player's health is scrutinized—before a game, during a game, and after a game. When a player's injury is a concussion, all of the one-off concussions of the past seem to come together, and the experiences of Marc Savard, Keith Primeau, Steve, and others suddenly matter. The need to play, versus the need to heal. The consequences of now, versus the consequences of the future. Everything is suddenly complicated. Everything consequential. Day after day.

Clarke MacArthur, an undersized, heart-and-soul left winger from Lloydminster, Alberta had played 548 NHL games, including one season with Steve in Buffalo, but only four games since February 16, 2015 when, while chasing down Jay McClement of the Carolina Hurricanes, the two of them crashed into Ottawa Senators goalie Robin Lehner, MacArthur striking Lehner's right shoulder face-first. It happened the day after Steve died.

MacArthur came back to play the following September, but during a pre-season game, after minor contact with teammate, Mark Fraser, his concussion symptoms returned. The symptoms were less severe this time, and he was soon back playing. But three weeks later, in a regular season game against Columbus, another seemingly innocent collision put him out of action the rest of the year.

He had eleven months to heal. Then in a training camp scrimmage in September, 2016, he was caught unawares by a high, hard, but not shuddering hit from Patrick Sieloff. Again, MacArthur was shut down. He took every treatment, followed every protocol, and tried as hard off the ice as he always had on it to get himself back playing, maybe even harder this time because he was thirty-one years old and knew this might be his last chance. Harder too because he had just signed a big five-year $23.25m contract and was embarrassed not to earn it and not help his already injury-depleted team fight for a playoff spot in the tightly competitive Eastern Conference.

By January, 2017, MacArthur was feeling a little better. He was training harder, and his teammates and coaches began to believe, as he did, that he might just make it back. And what a difference that would make if he could, late in the season, in the midst of a playoff race, being the kind of player and guy he is. In mid-January, he took a baseline concussion test that he and everyone knew he would pass. But he failed. The team announced that he wouldn't play for the rest of the year. His career seemed over.

But MacArthur kept at it—working out and skating day after day by himself. He knew that as good as he might begin to feel, he wasn't okay unless his baseline tests said he was. "So I studied up for those tests," he said later, "and did . . . all the things you can possibly do to be ready for [them]." And everyone knew that's what he would do, because that's who Clarke MacArthur is. He works hard in the gym, hard in practice, and if he wants something he will do what it takes to achieve it. For MacArthur, this was like a Stanley Cup game itself. The test was his opponent. What were its strengths? Its weaknesses? He had to beat it.

Then, near the end of the season, what was never going to happen, happened. He passed the tests, and on April 4, after the Senators had lost five straight games, he skated out onto the ice and played his first game in eighteen months. It's hard to know who was the most emotional about his return—his teammates, his coaches, his wife, Jessica, his parents, the fans, or MacArthur himself. Even the media. It was so right. So perfectly just.

Then things got better.

In the first round of the playoffs, the Senators were leading the Bruins three games to two. The sixth game was in Boston. MacArthur scored the game-winning, series-winning goal in overtime.

It was beyond belief. So thrilling, so life-affirming, it didn't matter if you hated the Senators or hated hockey and everything about it. It didn't matter if you were a player, or coach, or GM, or owner, or fan, or journalist, or a doctor and knew something about concussions and felt your insides empty every time you saw

MacArthur on the ice. It was such a great story. How good he must feel. How proud he must be. "It's ecstasy that you can't get anywhere else," MacArthur told *Toronto Star* reporter Bruce Arthur a few days later. "Scoring in OT, scoring the game-winner, I always say it's got to be like a 6/49 lottery knocking on your door and saying you've won $50 million." But even better than the lottery, he had done it himself, through his own sheer will.

And this was exactly how it should be. How life should be. Four concussions in eighteen months tells us a lot. It is the writing that is unmissably on the wall. But it doesn't tell us everything because we don't know everything. Sometimes things happen that we don't understand. We have to allow for that. We have to believe that we can create our own destiny every bit as much as science creates our destiny for us. After all, the concussion tests only say "not now." They don't say "not ever."

Once MacArthur had started things in motion, once he kept on trying, kept on training, kept on showing up at the rink, once he had failed his baseline test but wouldn't go home, what was going to stop him? Once he kept feeling a little better, and a little better, it was clear the team needed him and he needed the team, and the playoffs were getting closer. Once that mountain of hope and need to play built and built, higher and higher, and the story got better and better, who was going to say no to him? Who was going to tell him he couldn't play? MacArthur, his wife, his parents, his teammates, his coaches, his GM, his owner, the doctors, the NHL's Department of Player Safety, Gary Bettman? That would be so unconscionably cruel. He just wanted to feel normal again, and as it was with Marc Savard, normal was to play.

In Game 2 of the second round of the playoffs, Ryan McDonagh of the Rangers struck MacArthur with a high, hard, but not shuddering check. He left the game with what Ottawa called an "upper body injury," what they and MacArthur later said was a pinched nerve in his neck, unrelated to the previous concussions he had suffered. Clarke MacArthur returned to the lineup for Game 3.

He played every game for the rest of the playoffs, nineteen in all, as the Senators beat the Rangers in six games. MacArthur recorded two assists in the series-winning game. The team lost to the Penguins in the next round, in the seventh game in overtime. He had three goals and six assists in all, seventh most on the Ottawa team, and averaged over fifteen minutes of ice time a game, seventh highest among the Senators forwards.

A few days after the season was over, MacArthur talked with *Ottawa Citizen* reporter Ken Warren about all the amazing ups and downs of the year for the team and for him. "It's great to be back," he said. "As much as I wanted it to happen, I never dreamed it would happen—let alone about how far our team got." As for next season, suddenly circumspect, he said, "I just want to take a week or two and see how I feel. I still love playing the game." He told Warren how, now more than three weeks after his initial injury, he was still feeling discomfort in his neck. "I feel pretty good," he said. "I just want to make sure I'm all sorted out. It's my neck, not anything else. It should be good. From the first round [against Boston] I could feel it. I will get an MRI and go from there."

Hope, need to play, possibility, dream, excitement? The brain, as Dr. Lili-Naz Hazrati said, will tell its own story.

Then there was Sidney Crosby, Steve's buddy from Vail. In a career filled with improbable moments, the 2016-17 playoffs may have been Crosby's most improbable. Two days before MacArthur suffered his injury, Crosby, coming off the left wing, got a pass from team-mate Jake Guentzel and cut to the net. Washington Capitals' star Alex Ovechkin closed on him, cross-checked him across the shoulders, his stick glancing upwards, striking Crosby on the top of his helmet. Off balance, Crosby fell towards Capitals defenceman Matt Niskanen, who raised up his stick in front of him in cross-check position, hitting Crosby in the face. Play went on; Crosby remained on the ice.

Seven months earlier, Crosby collided in a pre-season practice with a teammate, felt his concussion symptoms come back, and missed the first six games of the year. It had been five years since his last concussion. In the time between, Crosby had won NHL trophies for most points, most goals, and MVP during the regular season and playoffs. He had captained the Penguins and Team Canada to a Stanley Cup, Olympic gold medal, World Championships, and a World Cup of Hockey Championship (where he was also named the MVP). He had become universally acknowledged (again) as the best player in the world. Then came that nothing moment in that nothing practice.

A story about Crosby in the *Toronto Star* a few days after his injury began, "We kind of hoped we were past this." Based on what? According to whom?

Then Crosby returned to play, and had one of the best of his many outstanding seasons. He led the league in goals, was second in scoring, and second in voting for the league's MVP. And the Penguins, ahead 2–0 in their series against the Capitals, the NHL's regular season champion, were suddenly a favourite to win another Cup. Now this.

The debate started up again. Hockey's critics criticized. Its defenders defended. It sounded like Marek, Shannon, Morrison, and the dead parrot all over again. Niskanen raised up his stick. It was intentional. No, it was accidental. Niskanen should be suspended. No, he shouldn't. Ovechkin's vicious hit had started things. No, it didn't. It was a "hockey play." No, it wasn't. Niskanen received a game misconduct, but was not suspended for any further games. Intentional? Accidental? The brain doesn't distinguish. The next day, the Penguins announced that Crosby had suffered a concussion. His playing status was termed "day-to-day." He did not play in Game 4.

Commentary about Crosby and his concussions that had been put on the shelf for five years, came down off it again. For his part, Crosby did what he had learned to do. Immediately after his injury, he made no comment, left the public eye, and retreated into a world of the few he trusts most, and mostly into himself. As he said later, "I know my body," and he needed quiet to know what his body was telling him. Three days after his injury he appeared again for practice. The next day, he passed his baseline test and engaged in full practice, which included the usual incidental contact. Hockey commentators assumed, at best, he wouldn't dress for Game 5 a day later. The Penguins were ahead three games to one in the series; if Crosby had any thought of playing again in the playoffs why not give himself a few more days to heal. It only made sense. That night, he played more than nineteen minutes, about his playoff average, got an assist on a goal by Phil Kessel, and played in the hard areas of the ice, in the corners and in front of the net, in the way he always does it.

The Penguins lost. In the first period of Game 6, Crosby was hit in the nose with a stick. In the second period, going to the net, he was knocked out of control and flew crumpled into the boards striking his shoulder and head. In a single instant, 19,000 Penguins fans sucked the air out of the arena. Crosby got up and continued. At the end of the game, *Hockey Night in Canada* host Ron Maclean said in summary that how Crosby had dealt with this tough, punishing night "will stamp out a lot of conjecture. . . . It's done," MacLean said, then as if hearing his own voice and becoming suddenly aware of everything that had come before for Crosby and other players, he caught himself and said, "Maybe it's not done." For games afterwards, when Crosby went onto the ice a confusion of feelings went with him—about him and his play; about him and his health. About hope; and about fear.

Crosby's play through the playoffs, in wins over Columbus, Washington, and Ottawa, had seemed more up and down than usual, and in speculation about the Conn Smythe Trophy as the playoffs' most valuable player, he was rarely if ever mentioned. That changed in the last few games of the finals against the Nashville Predators. Tied 2–2, Game 5 would turn the series one way or the other. And Crosby dominated. He dominated in every way—in what he did and how he did it; in making his teammates better and his opponents worse. The Penguins won 6–0. A few nights later, they won the sixth game and their second straight Stanley Cup, and Crosby his second straight Conn Smythe Trophy.

Had Crosby been affected by his concussion? He said nothing about it. To many, it seemed rash that he had come back so soon after the injury. He doesn't play a safe style of game. Gretzky and Mario Lemieux both gave themselves space with their ability to imagine the possibilities of a play, and then their ability to create it—Gretzky with his passes, Lemieux with his puck-handling and reach. Crosby, in today's faster, more congested game, has always had to fight for his space. His is a game of skill and will. He is the most Canadian of hockey's great stars. By continuing to play, he couldn't add much to what is already in the record books. He might only add to his legend: *remember the year Crosby returned from his*

concussion and, mucking and scrapping, took his team on his back, and won.
But Crosby didn't keep playing to add to his legend. Legends play,
as others do, because their team and their teammates need them,
because the season is still there to be won, and because every
player has only a few Stanley Cups in him. If Crosby didn't quite
know that in 2009 when he was twenty-one and won for the first
time, at twenty-nine, he does now. Other teams improve, age
catches up, the salary cap spreads Cup-winning players to the four
winds of the league. Every year might be a player's last, best chance.

When concussions don't disappear into the distraction of a
long season, when they matter every day, every game, we can see
them for what they are. For two months, the 2017 playoffs offered an
unmissable view of the twisting, turning, haunting, complicated
state of concussions and the game. In the celebration and glow of
achievement that follows, the draft, and free agency, this can be
quickly forgotten. The question is now what it has always been:
how can we make better decisions about concussions? How do we
create the circumstances for Gary Bettman, where it is difficult for
him not to do what needs to be done, and easy for him to do it?

First, Gary Bettman and the league need to recognize that the
problem of brain injuries comes from the way we play. And how we
play in the future is not inevitable.

Bettman might use his clever lawyer's mind instead to get out of
whatever awkward box he has put himself into, and put himself into a
better one. He might say that science is crucial in informing us and
guiding us and that, in fact, he and the NHL will work to enhance
and speed up science by helping to fund its best and most promising
research. But he could also say: Whatever science says about the connection between hockey and brain diseases, too many of our players
are experiencing lingering effects from their injuries—whether to

their knees or hips, or to their heads—and we want to do something more about that. We are always looking to reduce injuries in our game. That's why we created the Department of Player Safety ("the first of its kind in professional sports"). So we are going to take a comprehensive look at our game, as every sport should, to ensure that it remains the great, exciting game to play and to watch, but to see how we can make it safer, too. And because of the public's interest, we will put particular focus on head injuries.

There is a phrase many of us love and often use because we want it to be true, and we know it *can* be true. Ever hopeful and unconquerable: "Where there's a will, there's a way." Yet, when facing deep-seated, complicated problems, the obverse of the phrase is more often true: Where there's a *way*, there's a *will*. Because after experiencing failure upon failure, the will grows tired, and committing our whole selves to something and failing, hurts too much. The will is not gone, it's busy scaling shorter mountains. It is a *will* that needs a *way*. It needs a first step, that leads to a next step, and to a next, each one possible, all of them together leading excitingly towards something big and doable. *Give me a way and I will give you a will. If you don't, I will change what I know I can change, and manage the rest.*

It is why awareness isn't the answer. Awareness speaks only to the will. It is why scientists aren't the answer and decision-makers are. For a decision-maker, a problem without an answer, a *what* without a *how*, a *will* without a *way*, can be ignored and denied. They can try to sell you something they can do with their eyes closed and sell it like crazy as if it's the real answer, the real how, the real way, but which speaks instead to the dimensions of the problem as *they* see it, and not of the problem itself.

Until now, it seems as if Gary Bettman and those around him do not believe that there is a way to make this game just as exciting to play and to watch, just as successful off the ice and on, but significantly, appropriately safer. No answer; no problem. No way; no will.

But there *is* a way. Bettman and others could bring together the most respected people in the sport, whoever they are—players, former players, coaches, former coaches, as well as doctors and researchers—from Canada, from the U.S., from wherever in the world, and pose to them the question, and have them answer: *How do we reduce the frequency and force of blows to the head?* These are people who know the game, who know what makes it exciting to play and to watch, who are capable every day of seeing it with fresh eyes: what it is, not what they already know it to be. People who understand that there are a hundred ways to play the game, and to play it successfully, excitingly, and to win. If two players race into a corner for the puck, they know it isn't about who gets there first, or who hits whom, but who controls the puck and makes the play. They know that there is more than one way to make the play successfully, and that some ways leave a player more vulnerable to a head hit. What, then, are the best ways to make that play successfully *and* safely?

These players and coaches will come up with good answers. They are among the most creative people on earth because they are among the most competitive people on earth. They are always looking for something new and better so they can win. In this case, as it relates to hits to the head, what is that something better? Put them around the same table to talk about this, and do it in public so everybody can watch them and listen to them and think along with them, and understand as they understand; and online, so people are able to add their own thoughts. So when the discussion is over, it's not really over. So people will continue to talk about it, and think about it. And so the media will, too. And parents. And other players and coaches, at all levels. Because if these actions are good enough for the best in the world to take, they're good enough for them. So that when games are played, if what is implemented isn't good enough, as with every game plan, they can adapt and make it better.

For Gary Bettman and the league, this is not just about acknowledging problems, it is about focusing on answers. It is a chance to turn what can feel like grim necessity into exciting possibility. A chance not to run away from a problem, but to run towards an answer. A way and a will together. A chance to resolve, not manage, something that is not fair, not right, and not necessary.

Two small changes—no hits to the head; no finishing your check—and one process: actions of a dimension consistent with the dimension of the problem; actions to achieve, not just do; actions that Gary Bettman can implement easily. Actions that leave him no reason not to.

Steve Montador and others are gone. But something good can come out of something very bad. 2017 is the NHL's one-hundredth anniversary. It is a natural time to look back, to see where we've been, to see where we are, to see where we're going. To celebrate, and to thank those who have made the game what it is. But the best kind of celebration always looks forward even as it looks back. The game of the past doesn't matter much unless the game of the present is strong; unless the imagined game of the future is even better. Maurice Richard and Gordie Howe gave meaning to Cyclone Taylor and Howie Morenz before them, and Bobby Orr and Wayne Gretzky to Richard and Howe, because each made the game better. It is the players who come *after* 2017 who will give meaning to the NHL's first hundred years. Gary Bettman and the league need to give those players the chance to make the game better still. To make 2017 an anniversary worth commemorating.

Gary Bettman's contract runs five more years. During that time, he could play out the string, make lots more money, and feel the daily assurance that comes from doing what he already knows how to do. Or he could take on some big, important things. He could use the considerable power and skills he has developed to make hockey the exciting, amazing, safer, better game it can be in the future. Gary Bettman doesn't need to be a tobacco baron or a climate change–denier fighting against the night. His biggest contribution is still ahead.

ACKNOWLEDGMENTS

I have many people to thank. My father, Murray Dryden, who introduced me to the game; my brother, Dave, who showed me the joy of play; our children, Sarah and Michael, who allowed me to be both their parent and playmate. My teammates and opponents, with whom I learned about dreaming, trying, winning and losing, retrying—discovering that there's always a way. A lifetime of lessons that seemed only to be fun.

I want to thank those who trusted me with Steve's story. His family—Paul and Donna, Chris and Lindsay. His friends, in hockey and out—Mike Keating, Steve Valiquette, Andy O'Brien, Nick Robinson, Jay Legault, Marty Gélinas, Daniel Carcillo, Hayley Wickenheiser, Gisele Bourgeois, Rhett Warrener, Jamie McLennan, Daniel Tkaczuk, Kevin O'Flaherty, Mike Gardner, Mary and Terry Babcock, Missy Holas, Chantelle Robidoux, Craig Button, Jim Playfair, Joe Nieuwendyk, Jim Donaldson, and many more. They know Steve matters. He got inside all of them. He is inside them still. I want to thank Keith Primeau and Marc Savard, who shared many of Steve's same experiences and who helped me understand what his life might be like had he lived. Also the doctors and researchers who didn't know Steve, but who know how much the brain matters—Charles Tator, Karen Johnston, Lili-Naz Hazrati, Alain Ptito, Robert Cantu, and others—who want to see fewer patients, and who want much better lives for those they do see.

I want to thank those who helped make this book happen. Joe Lee and my editor at Signal/McClelland & Stewart, Doug Pepper, and Bruce Westwood at Westwood Creative Artists, all of whom understood the purposes, ambitions, and hopes of the book, and jumped in with both feet. My copy editor, Gemma Wain, who picked up things I had stopped seeing and made saves I could only dream of making. Also Evelyn Armstrong, who transcribed endless hours of interviews, somehow knowing when to end one cluster of words and start a new one.

I want to thank those, too, who were unfailingly encouraging—Dave Dryden, Roy MacGregor, Doug Gibson, John Macfarlane, Dan Diamond, and others—who had to suffer through my early drafts, pretending that what they were reading was worth encouragement. Their effect was beyond measure.

And my wife, Lynda. Lynda is always my first editor. She has in her hands that crucial, initial response—might what she is reading work, or not? Can it be what it needs to be, what I hope it will be, or not? With this book, I knew she would also be my most important editor. I asked her less to note problems and offer suggestions, and more to react to every paragraph she read. This book, above all, is about outrage and hope. Outrage for what is wrong; hope for what can be made right, and how. Lynda, in all of her life, has never for a moment lost, or had diminished, her sense of outrage and hope.

And one more person, whom I don't know and only saw once. It was about four years ago. He was about eight at the time, our grandson, Hunter, was six, and it was at a pre-season hockey tryout in Connecticut. There were probably thirty kids on the ice, but he was the one who jumped out. He could fly. He could do anything he wanted on his skates. Go in any direction, at any speed, make things up as he was doing them; he looked as if he were having the time of his life. And I was having the time of my life watching him, almost right away imagining what he might do next, what he'd be like at ten or twelve, wondering if he'd grow. And worried, too, almost right away, about what would happen to him when he got hit hard the first time. Not so much the injury he might receive, but whether that would take some of the joy out of him, out of his body, whether he would look different. Skate differently. And also, almost right away, about what would happen to him after his first concussion, because he would almost surely have at least one. What would that do to his brain? To him?

As I was writing this book, he was never not in my head. I hope for him.

Ken Dryden was a goalie for the Montreal Canadiens in the 1970s, during which time the team won six Stanley Cups. He also played for Team Canada in the 1972 Summit Series. He has been inducted into the Hockey Hall of Fame and the Canadian Sports Hall of Fame. He is a former federal member of parliament and cabinet minister, and is the author of five books, including *The Game* and *Home Game* (with Roy MacGregor). He and his wife, Lynda, live in Toronto and have two children and four grandchildren.